T0179085

S_{oy}
$A_{pplications}$
$_{in}F_{ood}$

Soy
Applications
^{in}Food

Mian N. Riaz, Ph.D.

CRC Press
Taylor & Francis Group
Boca Raton London New York

CRC Press is an imprint of the
Taylor & Francis Group, an **informa** business
A TAYLOR & FRANCIS BOOK

CRC Press
Taylor & Francis Group
6000 Broken Sound Parkway NW, Suite 300
Boca Raton, FL 33487-2742

First issued in paperback 2019

ISBN-13: 978-0-8493-2981-4 (hbk)
ISBN-13: 978-0-367-39170-6 (pbk)

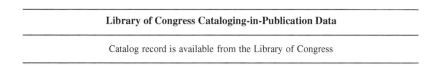

Library of Congress Cataloging-in-Publication Data

Catalog record is available from the Library of Congress

Visit the Taylor & Francis Web site at
http://www.taylorandfrancis.com

and the CRC Press Web site at
http://www.crcpress.com

Preface

Soyfood utilization around the world varies widely. Asia utilizes soybeans primarily as traditional foods such as tofu, soymilk, and fermented products, whereas Western nations consume more soybeans in the form of refined soy protein ingredients used in food processing than as tofu or soymilk. The consumption in Asia is based on longstanding traditional eating habits and food production methods. In Western nations, the consumption of soybeans as direct human food is a somewhat new phenomenon, although it is gaining increased acceptance and significance. Soybeans in food applications became very popular after the Food and Drug Administration approved a soy protein health claim in 1999. The use of soy in various food applications is of major importance to food industries. Because soy ingredients are being applied in so many diverse food systems, they are increasingly being regarded as versatile ingredients. Not only are soybean ingredients healthy, but they also play a major role in food functionality.

Currently, a limited amount of information is available on the use of soybeans in food applications, even though many types of soy ingredients can be used in various food systems. Sometimes it is difficult to decide what types of soy ingredients and processing will be best for certain food applications. This book provides insights into the different types of soy ingredients and their processing requirements for food applications. It serves as a source of information to all who are involved with the production of foods containing soy ingredients. For readers new to this area, the book can further their understanding of soy ingredients and their many applications.

This book summarizes some of the fundamentals to be considered when applying soy ingredients to food systems. The text is an excellent starting point for research and development personnel, students, and food technologists and other professionals in the food processing field. It brings together in-depth knowledge of processing food with soy ingredients and practical experience in the application of soy in food. It offers a wealth of information about the health benefits of soy protein, the current soyfood market, and the processing of soybeans into different soy ingredients. Also discussed is the use of soy protein in baked goods, pasta, cereal, meat products, and food bars. Some of the material addresses how to process soybeans into soy milk, soy beverages, and texturized soy protein; how to select identity-preserved soybeans for various food applications; how to overcome the beany flavor of some of the soy products; and how soy protein is fulfilling the need for protein in underdeveloped countries. This book is a valuable resource for information on the technical and practical applications of soy ingredients and will be a useful guide for selecting the proper soy ingredients for various applications. Most of the contributors to this book have at least 10 to 15 years of practical experience in their respective fields. The editor owes a large debt of gratitude to the many individuals who have provided information and inspiration.

The Editor

Mian N. Riaz, Ph.D., earned his bachelors and masters degrees in food technology from the University of Agriculture, Faisalabad, Pakistan, and his doctorate from the University of Maine, Orono. He is currently Head of the Extrusion Technology Program at the Food Protein Research and Development Center and a graduate faculty member in the Food Science and Technology Program at Texas A&M University, College Station. Dr. Riaz conducts research on extruded snacks, texturized vegetable protein, pet food, aquaculture feed, oilseed processing, and extrusion-expelling of oilseeds. He is also the editor of the book *Extruders in Food Applications* (Technomic) and co-author of the book *Halal Food Production* (CRC Press).

The Contributors

B. Bater
The Solae Company LLC
St. Louis, Missouri

M. Hikmet Boyacioglu
Food Engineering Department
Istanbul Technical University
Istanbul, Turkey

H. Chu
The Solae Company LLC
St. Louis, Missouri

Lynn Clarkson
Clarkson Grain Company, Inc.
Cerro Gordo, Illinois

Ignace Debruyne
Ignace Debruyne & Associates
Izegem, Belgium

Peter Golbitz
Soyatech, Inc.
Bar Harbor, Maine

Jim Hershey
American Soybean Association/
    World Initiative for Soy in Human
    Health (WISHH) Program
St. Louis, Missouri

Xiaolin L. Huang
The Solae Company LLC
St. Louis, Missouri

Joe Jordan
Soyatech, Inc.
Bar Harbor, Maine

Rongrong Li
Givaudan Flavors Corporation
Cincinnati, Ohio

Frank A. Manthey
Department of Plant Sciences
North Dakota State University
Fargo, North Dakota

M.K. McMindes
The Solae Company LLC
St. Louis, Missouri

Mark Messina
Loma Linda University
Loma Linda, California

I.N. Mueller
The Solae Company LLC
St. Louis, Missouri

A.L. Orcutt
Astaris LLC
St. Louis, Missouri

M.W. Orcutt
The Solae Company LLC
St. Louis, Missouri

Paul V. Paulsen
The Solae Company LLC
St. Louis, Missouri

Mian N. Riaz
Extrusion Technology Program
Texas A&M University
College Station, Texas

Brad Strahm
The XIM Group LLC
Sabetha, Kansas

Wesley Twombly
Nuvex Ingredients, Inc.
Blue Earth, Minnesota

Steven A. Taillie
The Solae Company LLC
St. Louis, Missouri

David Welsby
The Solae Company LLC
St. Louis, Missouri

Contents

1

Soyfoods: Market and Products

Peter Golbitz and Joe Jordan

CONTENTS

It can be eaten raw, roasted, fermented, or cultured. It can be a small, green bean or a magically transformed concoction that closely resembles hamburger, ice cream, bacon, or even a fully dressed turkey. Yes, the soybean is one of the most nutritious and versatile foods on the planet, and it is an important food platform for the 21st century.

History of Soyfoods

Even though soy is a new food for many in Western society, the Chinese have considered it an important source of nutrition for almost 5000 years. The first reference to soybeans, in a list of Chinese plants, dates back to 2853 B.C., and ancient writings repeatedly refer to it as one of the five sacred grains essential to Chinese civilization. The use of the soybean in food spread throughout the Asian continent during the early part of the last millennium, as people in each region developed their own unique soyfoods based on tradition, climate, and local taste preferences. Natto, for instance, is a product consisting of fermented soybeans that was developed at least 3000 years ago in Japan and continues to be popular in some regions of that country today.

Growth and Development in the Western World

When European missionaries and traders traveled to Asia during the 1600s and 1700s, they wrote in their journals about traditional soyfoods, such as tofu and soymilk, that they had encountered in the cultures they explored. Then, in the late 1700s, both Benjamin Franklin and a curious world traveler named Samuel Bowen sent soybean samples to the United States for cultivation. Mr. Bowen's soybean crops, grown in Georgia, were even used to manufacture soy sauce and a soy-based vermicelli substitute. But, it was not until Asians began to emigrate to Europe and North America during the 1800s that soyfoods were consistently made by and for people in the United States. Several Chinese tofu and soymilk shops were established in cities with large Asian populations in Europe and on the East and West coasts of the United States; however, throughout the 19th century, soyfoods tended to be made in small, family-run shops and were distributed and consumed primarily in Asian neighborhoods. During the 1920s, a number of smaller companies with ties to the Seventh-Day Adventist Church (many members of which are vegetarian) began making tofu in the states of Tennessee and California. During that same time, soy flour started to gain favor in both Europe and the United States as a low-cost source of protein in the production of meat substitutes, and, during both World Wars, large amounts of soy flour were used to help offset meat shortages.

Soybean Industry Blossoms in the United States

Large-scale development of the soybean crop and processing industry began in the United States during the 1940s and 1950s, spurred on by a rapid increase in domestic and worldwide demand for both protein meal and vegetable oil. Harvested soybean acreage in the United States more than tripled between 1940 and 1955, from 4.8 million acres to 18.6 million, while total production of soybeans increased nearly fivefold, from 78 million bushels to 374 million. As the number of acres devoted to soybeans continued to grow during the 1960s, the United States became a world soybean superpower and began exporting large quantities of soybeans, as well as the basic crush products of meal and oil, to Europe and Asia. By 1970, U.S. farmers had planted an incredible 43 million acres of soybeans and produced 1.1 billion bushels. The crop would more than double in size to 2.3 billion bushels, or 62.5 million metric tons, by 1979. Industry growth has slowed in recent years due to increased competition from other producing nations, but the United States still produces roughly 75 million metric tons of soybeans each year.

Soybeans Grow Around the World

Though still the largest soybean exporting country in early 2005, the United States has lost the dominant position it once had in the global soy trade. Brazil, Argentina, and India have all become major producers as the world's demand for soy as food, vegetable oil, and animal feed has continued to increase. Given the amount of available arable land and water resources in Brazil, along with its low labor costs, it is expected that Brazil will eventually become the number one soybean-producing nation. Already, South America as a continent produces more soybeans that North America (combined U.S. and Canadian production). Growth in China, where this story began, has been plagued with inefficiencies and lags behind most major producing countries, although it is still the fourth largest soybean grower. In 2004, it was the world's largest importer of soybeans as well.

TABLE 1.1

World Soybean Production by Major Producers

| | Production (million metric tons) | | | | |
	2000/01	2001/02	2002/03	2003/04	2004/05
United States	75.06	78.67	75.01	66.78	85.48
Brazil	39.50	43.50	52.00	52.60	63.00
Argentina	27.80	30.00	35.50	34.00	39.00
China	15.40	15.41	16.51	15.40	18.00
India	5.25	5.40	4.00	6.80	6.50
Paraguay	3.50	3.55	4.50	4.00	5.00
All others	9.42	8.60	9.60	10.23	11.64
Total	175.93	185.12	197.12	189.81	228.63

Source: U.S. Department of Agriculture estimates, March 2005.

Soybean Production and Utilization for Food

For the marketing year 2004/05, the world produced approximately 229 million metric tons of soybeans, enough to give each man, woman, and child 35 kg of soybeans each, or the equivalent of nearly 300 L of soymilk for a year (see Table 1.1). But, the world's soybeans are not used exclusively for humans. Each year, on average, an estimated 85% of the world's soybeans are processed (crushed) into soybean meal (used primarily for animal feed) and vegetable oil. Approximately 10% is used directly for human food and the balance is used for seed or on-farm feeding of animals or is waste material. In addition to the 10% of the crop that is used directly for human food — for products such as tofu, soymilk, natto, and miso — an estimated 4 to 5% of the soybean meal that is produced from soybean crushing is further processed into soy protein ingredients that, in turn, are further processed into various meat and food products or are used in infant formula, dairy and meat alternatives, nutritional supplements, and energy bars.

Soyfood consumption patterns around the world vary widely, with such regions as Asia utilizing soybeans primarily in such traditional foods such as tofu, soymilk, and fermented products. In Western nations, more soybeans are consumed in the form of refined soy protein ingredients (used in food processing) than in the production of tofu and soymilk. As well, Soyatech, Inc. (Bar Harbor, ME) estimates that Asian nations utilize 95% of the soybeans consumed directly as human food in the world today. The heavy consumption in Asia is based on long-standing traditional eating patterns and food production methods. In Western nations, consumption of soybeans as direct human food is a somewhat new phenomenon, although it is gaining increased acceptance and significance.

As can be seen in Table 1.2, Asian countries consume far more soybeans and soy protein equivalent per day than do Western nations. For example,

TABLE 1.2

Annual Per Capita Consumption (2001) of Soybeans for Direct Food

Rank	Country	Soybeans (kg/yr)	Soybeans (g/day)	Soy Protein Equivalent (36% Protein) (g/day)
1	Taiwan[a]	19.15	52.46	18.89
2	Korea (North)	10.67	29.24	10.53
3	Korea (South)	8.79	24.07	8.67
4	Libya	8.68	23.78	8.56
5	Japan	7.73	21.19	7.63
6	China	7.31	20.03	7.21
7	Indonesia	7.16	19.61	7.06
8	Uganda	4.71	12.91	4.65
9	Nigeria	2.76	7.57	2.72
10	Thailand	2.34	6.40	2.30
11	Myanmar	1.91	5.22	1.88
12	Yemen	1.85	5.06	1.82
13	Costa Rica	1.40	3.84	1.38
14	Peru	1.40	3.83	1.38
15	Vietnam	1.27	3.48	1.25
16	Canada	0.68	1.88	0.68
17	Zimbabwe	0.65	1.77	0.64
18	Philippines	0.51	1.39	0.50
19	India	0.41	1.13	0.41
20	Ethiopia	0.38	1.04	0.38
21	United States[a]	0.33	0.89	0.32
22	Germany	0.24	0.66	0.24
23	Egypt	0.24	0.66	0.24
24	South Africa	0.23	0.64	0.23
25	Congo, Democratic Republic	0.18	0.50	0.18
World average		2.39	6.54	2.36

[a] Soyatech, Inc., estimates.
Source: Food and Agriculture Organization (FAO) food balance sheets.

in Taiwan, per capita consumption is estimated to be as high as 19.15 kg per year, and in Japan at 7.73 kg per year. In the United States, yearly per capita consumption of soybean is just 0.33 kg per year; however, world average per capita consumption is approximately 2.4 kg per year, equivalent to around 6.5 g of soybeans per day, or an estimated 2.4 g of soy protein per person per year.

Soyfoods in Asia

Throughout Asia, the soybean is used in a wide variety of traditional and modern food products. For example, in Japan, tofu is the most popular soy-food consumed, eaten at virtually all meals in one form or another, from silken tofu in miso soup for breakfast to plain or fried tofu for lunch or dinner. Also, tofu is used in dessert products and as an ingredient in cutlets and other prepared foods. Soymilk, after a quick rise and fall in sales in the mid-1980s, is making a strong comeback in Japan due to increased interest in functional foods and beverages and recognition of the powerful nutritive qualities of the soybean. Natto, a fermented soyfood product, has historically been extremely popular in Japan and is consumed daily by many people. Natto is prized for its unique taste and form, as well as its powerful blood-thinning and cleansing qualities. Miso, another fermented soyfood, is consumed daily by many Japanese in soup broth, salad dressings, and food toppings.

In China, tofu is also very popular, as is fermented tofu, yuba (dried soymilk skin), soymilk, and a variety of regional specialties, including soy noodles. In addition, soy powder mixes and isolated soy proteins are becoming popular as food ingredients and in consumer-oriented mixes for health. In Taiwan, the art of meat substitutes and alternatives has reached new heights with meat-, chicken-, and fish-like products made from soy proteins, yuba, gluten, and tofu. In Indonesia, tempeh is the most popular soyfood and is sold at thousands of food stands and kiosks throughout the country. It is made fresh each day by many individuals who purchase inoculated soybeans the night before and by morning have fresh tempeh to sell as street vendors. Throughout Asia, in addition to tofu and other regional soyfoods, processed and packaged soymilk has grown increasingly important in recent years and has become big business in Hong Kong, Korea, Malaysia, Singapore, Thailand, and Vietnam.

Soyfoods in Europe

European consumption of soyfoods is similar to that of the United States, with meat and dairy alternatives comprising most of the soyfood sales. Soymilk and meat alternatives sales are particularly strong in the United Kingdom, which has a relatively large vegetarian population. Wide acceptance of soymilk can also be found in Belgium, where the continent's largest soymilk producer is located. Throughout the rest of Europe, tofu is known, but it is not as popular as meat and dairy alternatives. As in the United States, soyfoods have become more of a mainstream food item, having crossed over from the natural products market to being widely available now in mainstream supermarkets.

Soyfoods in Africa

While a pan-African consumer market for soyfoods has yet to develop, certain countries in Africa have readily adopted soyfoods due to their high protein level and nutritional quality. Some of the use is a result of food aid programs where soybean products such as textured soy flour or soybean–corn meal are distributed, while in some countries feeding programs for workers and school children take advantage of the relatively low cost and high nutritive value of the soybean. In South Africa, a modern soyfoods market has developed; fresh and aseptically packaged soymilk is available, as well as a wide array of products made from textured soy flour and sold as soy "mince," a low-cost ground meat replacement.

Soyfoods in the United States

In the United States, perhaps one of the more interesting markets for soy products has developed due to a number of supportive micro and macro trends, as well as a strong history of "Americanization" of foreign and ethnic foods. This is due to the United States being a large, immigrant melting pot that blends many different foods and eating traditions. In addition, a strong entrepreneurial spirit exists among both immigrants and those who are motivated by a desire to create livelihoods that benefit not only themselves but also the world at large. Soyfoods, due to their high nutritional value and their low environmental impact when compared to meat and dairy production, have become a rising star in the U.S. food marketplace.

Development of the U.S. Soyfoods Industry

The modern soyfoods industry in the United States has developed in three distinct phases. The first phase, one of *discovery*, took place during the 1920s, when early proponents of vegetarian diets such as Dr. John Harvey Kellogg of Battle Creek, MI, became interested in the healthful properties of the soybean. In addition to studying the usefulness of soybeans in the diets of diabetics, he also developed and marketed North America's first meat and dairy analogs made from the soybean. During those early years, soyfoods were promoted primarily among members of the largely vegetarian Seventh-Day Adventist Church and other special interest groups.

The second phase was distinctly *industrial* and paralleled the industrialization of American society. As mentioned earlier, soy flour had become an important component in food production during the Second World War, as meat shortages developed due to the increased protein and food needs of the armed forces and the disruption of farm commerce during the war years. Unfortunately, soy flour and protein processing was not a well-developed science at that time, and, as a result, most of the products had distinctive off-flavors and were tolerated at best, but not appreciated. The image of soybean-based foods became that of an inferior substitute and extender, to be used in times of crisis and shortage, rather than that of a nutritional alternative to meat and dairy foods for everyday use.

In the 1950s and 1960s and during the meat shortages of the 1970s in the United States, numerous food companies and meat processors used textured soy proteins to extend meat products. These early attempts were met with resistance by consumers who complained of poor flavor, texture, and color. Following this period, many food processors found it necessary to reassure consumers that their products contained no fillers or cereal additives. As this industrial phase ended, soybean-based foods were neither highly respected nor desired by American consumers.

It was during the 1970s that a third phase for soy emerged, a *new age*. In both the United States and Europe, a large, young counterculture had developed that began to question traditional American and Eurocentric values based on industrialization, military interventionism, and the politics of a meat-centered diet. New Age thinkers popularized the concepts of going back to the land, pacifism, vegetarianism, and the more equitable distribution of food resources. Two important publications of that era were *Diet for a Small Planet*, in which Frances Moore Lappé wrote of the misallocation of food resources and the value of the soybean, and the *Book of Tofu*, in which William Shurtleff and his wife Akiko Aoyagi wrote of, and eloquently illustrated, the beauty and tradition of soyfoods. The rediscovery of the value that the soybean could have in a modern world helped to inspire a new soyfoods movement in both the United States and Europe. During this time, hundreds of small companies were founded that were dedicated to producing soyfoods and educating the public about their use. The desire for a "right livelihood" within these companies and among their customers inspired tremendous innovation and helped to build a natural food and products industry and market.

According to statistics gathered by Soyatech, Inc., over 2000 new soyfood products were introduced during the 1980s in the United States, and many of them were made by small companies that made primarily (or exclusively) soy-based foods. Products such as tofu, tempeh, miso, tofu hot dogs, veggie burgers, tofu ice cream, soymilk, and other dairy alternatives all became common fixtures in natural food stores. As the young counterculture that founded this movement began forming households of their own, these soyfoods progressively crossed over into mainstream supermarkets.

The Americanization of Soyfoods

The pace of innovation and new product development accelerated throughout the 1990s with the continued Americanization of soyfoods into products that were more familiar and convenient, as well as healthful. In particular, the 1996 decision by White Wave, Inc. (Boulder, CO), to market its soymilk in traditional "gable-top" milk cartons in the refrigerated food cases of supermarkets led to a dramatic shift in soymilk consumption. Until that point, most soymilk was sold in natural foods stores and was packaged in aseptic packaging. Primarily vegetarians, the lactose-intolerant, and people who had ethical or religious objections to drinking cow's milk consumed it. Later, however, when White Wave positioned its Silk® brand soymilk to taste and look more like dairy milk, sales of soymilks in the United States exploded from $124 million in 1996 to nearly $700 million in 2004. Soymilk became simply another, perhaps more healthful, choice in the dairy case.

Another, broader impetus to the popularity of soyfoods was a 1999 decision by the U.S. Food and Drug Administration (FDA) to allow food manufacturers to include a claim for heart health on foods that contained more than 6.25 g of soy protein per serving (if they were also low in fat). This

TABLE 1.3

U.S. Soyfoods Market (1996 to 2005)

Year	$ (Millions)	Growth Rate (%)
1996	1244	11.5
1997	1484	19.3
1998	1747	17.8
1999	2288	31.0
2000	2769	21.0
2001	3234	16.8
2002	3648	12.8
2003	3912	7.2
2004	3996	2.1
2005 (projected)	4218	3.0

Source: Soyfoods: The U.S. Market 2005, Soyat-ech, Inc./SPINS, Inc., Bar Harbor, ME, 2005.

health claim states, "25 grams of soy protein a day, as part of a diet low in saturated fat and cholesterol, may reduce the risk of heart disease. A serving of [name of food] provides [amount] grams of soy protein." Since then, the range of foods that include soy protein has been limited only by the imagination of food scientists and marketers. In addition to soymilk and other dairy alternatives, categories that have done particularly well include tofu, energy bars, meal replacements, and meat alternatives. Categories on the rise include soy-based snacks, chips, soynuts, edamame (fresh green soybeans), and soy-enriched pasta, breads, and cereals.

The overall soyfoods industry has grown dramatically since the mid-1990s, from $1.2 billion in 1996 to an estimated $4.0 billion in 2004 (see Table 1.3). Though the industry is beginning to show signs of maturity and growth has slowed, soyfood sales are still outpacing the growth in the grocery market as a whole, with prepared convenience foods, snacks, and dairy and meat alternatives growing quickly. In the near future, the U.S. market can expect to see a wide assortment of new products, including snack foods, chips, cultured products, and a new generation of meat products that closely resemble muscle meat.

Soybean Nutritional Components

In comparison to many of today's major food sources, soybeans are truly a nutritional superpower. They contain the highest amount of protein of any grain or legume; substantial amounts of fat, carbohydrates, dietary fiber, vitamins, and minerals; and a veritable pharmacy of phytochemicals useful for the prevention and treatment of many chronic diseases. Soybeans vary widely in nutrient content based on the specific variety and growing conditions, but typically they contain 35 to 40% protein, 15 to 20% fat, 30% carbohydrates, 10 to 13% moisture, and around 5% minerals and ash.

Soy Protein

The protein in the soybean contains all eight amino acids essential for human health, but until recently it was accepted that soy protein was lower in quality than many animal proteins. These earlier assumptions were based on an older method of evaluating protein quality, the protein efficiency ratio (PER), which is based on the growth rates of rats as measured in laboratory tests. However, rats require 50% more methionine (one of the amino acids found in soybeans) than humans do, making this particular method inappropriate for evaluating soy protein quality for human consumption. In order to make up for the shortcomings of the PER evaluations, the World Health Organization of the United Nations and the U.S. Food and Drug Administration have adopted a new method for evaluating protein quality called the *protein digestibility corrected amino acid score* (PDCAAS). This method uses an amino acid score, a comparison between the amino acid pattern of a protein and human amino acid requirements, and a factor for digestibility to arrive at a value for the quality of a protein. Using the new PDCAAS method, soy protein products generally receive scores of between 0.95 and 1.00, the highest value possible.

Soy Oil

Soybeans, in comparison to other beans, grains, and cereals, contain a high amount of fat, but, as we have come to learn, all fats are not created equal. Many of our major health problems today are due to the fact that people eat too much fat, and the fat being consumed is unhealthful and of poor quality. Fortunately, the fat found naturally in the soybean — and, by extension, in most traditionally processed soyfoods such as tofu, soymilk, tempeh, full-fat soy flour, and liquid soybean oil — can be categorized as a healthful fat. Approximately 50% of the fat in soybeans is linoleic acid, an essential polyunsaturated fat that can help lower cholesterol by bringing down blood lipid levels. In addition, soybean oil can contain as much as 8% alpha-linolenic acid, which is an omega-3 fatty acid (the healthful fat commonly derived from fish) and which is believed to be beneficial in lowering the risk of heart disease.

Carbohydrates and Fiber

Soybeans contain an interesting mix of both soluble and insoluble carbohydrates (including dietary fiber) that together constitute about 30% of the soybean. The primary soluble carbohydrates in the soybean are sugars: stachyose, raffinose, and sucrose. Collectively, they make up about 10% of the soybean, although the amounts of these sugars vary according to the variety of soybean and its growing conditions. Raffinose and stachyose, the primary oligosaccharides (complex sugars) in soy, are significant because they are not digested or used as nutrients directly by the human body, but they are used as nutrients by the *Bifida* bacteria in the lower intestine. These types of intestinal flora are considered important for human health. It is believed that their presence can reduce the incidence of colon cancer and other intestinal disorders; however, when the bacteria break down these sugars, intestinal gas is created as a byproduct, creating discomfort and flatulence in some people. This can create a barrier to soybean consumption (especially in Western culture), but some Japanese companies are actually isolating and marketing these sugars as health supplements and food ingredients. The insoluble carbohydrates, or dietary fiber, come primarily from the outer hull and structural cell walls of the soybean and are composed of cellulose, hemicellulose, and pectin. This component contributes to the overall healthfulness of the soybean, because consumption of adequate amounts of dietary fiber has been shown to reduce the risk of heart disease and cancer, as well help to improve bowel function.

Vitamins and Minerals

In addition to providing high-quality protein, fat, and carbohydrates, soybeans are also rich in vitamins, minerals, and a number of other valuable phytochemicals. The major mineral components of soybeans are potassium, sodium, calcium, magnesium, sulfur, and phosphorus. Mineral content can vary widely due to both the type of soil and growing conditions for the soybean. Although soybeans are not considered to be very rich sources of any one particular vitamin, they do contribute to overall nutritional well-being. The water-soluble vitamins in soybeans are thiamine, riboflavin, niacin, pantothenic acid, biotin, folic acid, inositol, and choline. An integral part of lecithin, choline has been linked to the health of cellular walls and the nervous system. In 2001, the FDA formally recognized the healthful properties of choline by agreeing to allow food manufacturers to add a choline health claim to product labels. Fat-soluble vitamins present in the soybean are vitamins A and E. Vitamin A exists as provitamin beta-carotene and is present in higher levels in the immature, green vegetable soybean than in the mature (dry) soybean. Tocopherols, the most widely available, naturally occurring vitamin E compound, fill two major roles as a component of soy oil. First, vitamin E is an important element of human nutrition, although its bioactive properties have been scrutinized in recent years, and many of the health claims about vitamin E remain unproven. Second, tocopherols are antioxidants, which means that they are chemicals that prevent a substance from reacting with oxygen. Their presence in soybean oil slows down the degradation of the oil.

Isoflavones

It should come as no surprise that plants contain powerful chemicals that can have a profound effect on an individual's health or well-being. Many of the drugs used in Western medicine today originate from plant sources. Traditional medicines, such as those used by some Asian cultures, herbalists, and homeopaths, are centered on plant remedies. The term *phytochemical* is used to describe a class of plant-based chemical compounds that have an effect on the human or animal organism. The soybean is a virtual pharmacy of beneficial phytochemicals for human health and disease regulation; however, over the past few years, one particular set of compounds has gained the most attention and become the focus of hundreds of studies and numerous world conferences — isoflavones. For all practical purposes, no other food contains as significant an amount of these chemicals as does the soybean. Generally speaking, minimally processed soyfoods, including full-fat soy flour, tofu, and soymilk, have the highest levels of isoflavones. Isoflavones are also considered phytoestrogens, or plant estrogens, because they have a similar chemical make-up and effect on the human body as estrogen; however, the estrogenic effects of soy isoflavones are much (perhaps as much as 10,000 times) weaker than the human estrogen hormone. The major isoflavones in soybeans are genistein, daidzein, and glycitein. Genistein has shown some promise in preventing and treating prostate and breast cancers. Although glycitein comprises only 5 to 10% of soy isoflavones, recent studies suggest that it may have far higher estrogenicity and bioavailability than either genistein or daidzein.

Soyfoods and Protein Ingredients

As indicated earlier, approximately 10% of the world's soybean crop is used directly for human food, and a stunning array of products is made from the humble bean. Many of the following soy-based foods utilize the whole soybean, while some are made with a variety of soy protein ingredients, including isolated soy protein, soy protein concentrate, and soy flour.

Whole Dry Soybeans

Whole dry soybeans are, of course, the original soyfood. Dried in the pod while still in the field, whole soybeans contain approximately 37% protein, 17% fat, 10% dietary fiber, 20% carbohydrates, 5% ash (total minerals), and 11% moisture. Composition varies among the many different varieties of soybeans. Some seeds are larger in size and higher in protein than others, while some varieties have a brown, buff, or clear-colored hilum (the spot on the soybean where it connects to the pod). Dried, yellow soybeans are the most commonly available type. Soyfood processors carefully select the proper varieties of soybean for the type of products they are making. For example, soymilk and tofu manufacturers prefer large, high-protein soybeans with clear hilums so their finished product yields are high, exhibit a mild flavor profile, and are light in color. Most identity-preserved food-grade varieties are sold at a premium to food processors. Dried soybeans will last over a year but must be kept cool and dry, as they begin to degrade when stored at too high a temperature or in moist conditions.

Tofu

Tofu is perhaps the most widely consumed soyfood in the world today. It is a regular part of the diet in many Asian nations. Tofu is widely available across the United States and in most other Western nations. This soft, white, almost cheese-like product is favored for its versatility, mild flavor, and high nutritional value. It is naturally processed and, as a result, retains a good deal of the important nutrients and phytochemicals of the soybean, such as the isoflavones. Tofu is especially valuable due to its chameleon-like quality of being able to take on the flavor of whatever spices and ingredients are used with its preparation. For example, in the same sitting, one could dine on a fresh green salad served with a creamy tofu dill dressing, eat a healthy serving of marinated barbecued tofu, and finish with a tofu chocolate cream pie.

When tofu is made, soaked whole soybeans are ground to produce a slurry, which is added to water and boiled. After cooking, the pulp is removed from the mixture, leaving soymilk. While the soymilk is still hot, a natural mineral coagulant — such as calcium sulfate or magnesium chloride, or a mixture of both — is slowly added to the hot liquid. Within minutes, the soymilk begins to curdle and large white clouds of tofu curd begin to form in a sea of yellow whey. After 15 minutes or so, the curds are removed from the whey and placed under pressure in cloth-lined forming boxes. The curds are then pressed to form soft, regular, firm, or extra-firm tofu. The size of the curd and length of pressing time determine the style of tofu produced. The softer the tofu, the lower the protein and fat levels and the higher the water content. Soft tofu is also usually smoother in texture than firm tofu. Firm tofu, on the other hand, is higher in protein and fat and lower in moisture and has a denser, chewier texture.

Silken tofu — the soft, smooth variety commonly sold in an aseptic package — is made in a slightly different manner. To make silken tofu, either calcium sulfate or glucono-delta-lactone is added to a thicker, richer soymilk, and the mixture is put into a package. This package, with the soymilk and coagulant mixture, is heated to the proper temperature to activate the coagulation, and the soymilk is transformed into one solid, smooth curd, right in the package.

Typically, tofu contains between 10 and 15% protein and 5 to 9% fat. It is relatively low in carbohydrates and in fiber (because the pulp was removed), making it easy to digest. Tofu made with calcium sulfate or calcium chloride contains higher levels of calcium than those made with other coagulants and is therefore sometimes sought after by those seeking to supplement their current calcium intake. One 4-oz. serving of calcium-coagulated tofu can contain as much bioavailable calcium as one 8-oz. serving of cow's milk.

Tofu is commonly found packed in sealed, water-filled tubs, but it is also available in vacuum or aseptic packaging. Unless it is aseptically packaged, tofu requires refrigeration at or under 40°F. Tofu can also be frozen for longer storage, although it will tend to have a much different texture when thawed, becoming crumbly and more chewy. Regular, pressed tofu is best when fried, baked, grilled, or barbecued; used as a meat alternative; or added to stir fry dishes. In most cases, the best way to prepare tofu for use is to first drain the tofu on paper or cloth towels to reduce the water content. This improves the ability of the tofu to absorb flavors, reduces the amount of water that has to be removed in the cooking stage, and firms up the tofu, making it easier to handle.

Silken tofu is best used in soups or blended into vegetable spread, sauces, cream substitutes, pie fillings, puddings, or desserts. Some firm varieties of silken tofu are available, as well, and these can be used as pressed tofu in many recipes. The Japanese favor silken tofu and usually eat it prepared very simply. For example, it may be served fresh, with just a little soy sauce and scallions, or in miso soup. Silken tofu, because it is soft and lacks the cohesiveness of pressed tofu, is more difficult to handle. Many new tofu products on the market today have been flavored and then baked, fried, or smoked prior to sale. These new tofu products are "recipe ready" and easier to use for many consumers, especially those with a limited amount of time to cook or those who are unfamiliar with tofu.

Soymilk

Traditionally, soymilk is the liquid extract of the soybean, which can be used in the preparation of tofu or as a nutritious beverage, but the beverage-quality soymilks available today are usually prepared in a slightly different fashion, utilizing a number of more modern food processing techniques in order to produce a blander product with greater appeal to Western tastes. As with tofu, soymilk generally contains most of the active phytochemicals present in soybeans, including high amounts of isoflavones. Some soymilks are made with isolated soy protein as a base. Many are fortified with vitamins and minerals (such as vitamins A and D and calcium) to bolster their position as a viable alternative to cow's milk. Soymilk can be used as a direct, one-to-one replacement for cow's milk in most food formulations and recipes. Soymilk works very well in baking recipes and is an excellent cream soup or sauce base. It can be put on cereal or made into yogurt, pudding, or ice cream. A variety of powdered soymilk products can be mixed with water at low ratios if one needs a very thick, cream-like soymilk base.

Tempeh

Tempeh is a traditional, fermented soyfood that is unique in its texture, flavor, and versatility. Originally from Indonesia, tempeh has a flavor that is distinctively different from other soyfoods, sometimes described as "nutty" or "mushroom like." It is made from the whole soybean, which has been dehulled, cracked, and cooked in water with added vinegar to reduce the pH. Once cooked, the soybean is mixed with the spores of the *Rhizopus oligosporus* fungus and left to incubate for 24 hours at 88°F. At the end of this period, the tempeh is a compact, cake-like product that is completely covered with, and penetrated by, white mold mycelia. Tempeh may also be made with other grains or seeds mixed in during processing to vary the taste and texture of the final product. Tempeh contains about 19% protein, is higher in fiber than tofu, and is a significant source of vitamins and minerals. Tempeh is available fresh or frozen and is sold in plain and flavored varieties. It can be found in natural food stores and in the produce section of many supermarkets. Tempeh should be used within a few weeks of purchase but can be frozen for longer storage. It can be marinated prior to use, works very well in stir-fry dishes, and can be baked, grilled, or deep fat fried. Tempeh can also be grated and formed into patties or balls.

Soymilk Yogurt

Soymilk yogurt is made in the same manner as cow's milk yogurt. Pasteurized soymilk is inoculated with *Acidophilus*, *Bifida*, or other suitable cultures and incubated until the culture has turned the soymilk into yogurt. It tastes very similar to cow's milk yogurt, is available in a variety of styles and flavors, and is generally sold in 6- to 8-oz. and quart containers. It is very high in protein and a great source of isoflavones. These products may not be labeled as "yogurt" in the United States, as they are not made with cow's milk, but manufacturers package them in familiar yogurt containers and refer to them as "cultured soy" or "cultured soymilk."

Miso

Miso is a rich and flavorful paste made from either fermented and aged whole soybeans or soybeans in combination with wheat, barley, or rice. This salty paste is a treasured soup base and flavoring ingredient used throughout Japan, Korea, Taiwan, Indonesia, and China. Many types of miso are available in the world today, from sweet white miso, which is quite mild, to dark savory miso, which is much more robust and salty. Miso has some unique medicinal properties and is believed to help reduce the effects of radiation and other environmental poisons on the body. It contains enzymes and bacteria that can aid in digestion. It is high in protein, but it also contains high levels of sodium and should be consumed sparingly. To produce miso, whole cooked soybeans are mixed together with koji nuggets — grain such as wheat, rice, or barley that has been cultured with a fungal starter (*Aspergillus oryzae*) — and fermented under very specific conditions for the type of miso being made. When the mash is fully ripened, it is blended and packaged for sale. Usually the longer the miso has aged, the more complex the flavor. Most of the miso sold today is pasteurized and refrigerated. Miso is sold in natural food stores and in many supermarkets along with other Asian foods. It has a long shelf life when refrigerated. Due to its distinctive fermented flavor, it can be added to recipes or blended with tofu to add a cheese-like note.

Soy Sauce

Soy sauce is the most well known and popular of the traditional soyfoods and is used extensively as a flavoring ingredient in most Asian cooking. When naturally processed, soy sauce is produced in a manner similar to that of miso. When made exclusively with soybeans, the product is called *tamari*. If it is processed with a fermented wheat starter, the product is called *shoyu*. Much of the soy sauce sold today is not naturally fermented. Instead, it is made with hydrolyzed vegetable protein (HVP), sugar, color, and preservatives. HVP is produced from soy protein using chemically induced fermentation. All soy sauces are high in sodium and should be used sparingly. Some reduced-sodium varieties are available on the market today, as well as a number of flavored soy sauce products. From a nutritional prospective, tamari contains the highest protein level of the soy sauces, followed by shoyu and then HVP-based soy sauce; however, the high amount of sodium in all of these products should preclude anyone from using soy sauce as a nutritional food. Naturally processed soy sauce is available in natural food stores; both the naturally fermented and HVP forms are sold in most supermarkets. Due to its high salt content, soy sauce has a long shelf life without refrigeration but will keep longer when stored at cooler temperatures.

Okara

Okara is the fibrous remains of the soybean after it has been processed to make soymilk. It is very high in moisture and contains the insoluble carbohydrates and dietary fiber of the soybean, as well some remaining protein and fat. When fully cooked, it has a bland flavor and makes an excellent addition to breads and other baked goods. It can also be used to make meat alternatives or processed into tempeh. It is not usually sold in stores, as it is very wet, heavy, and highly perishable.

Natto

Natto is a whole soybean product, popular in Japan, that is produced by fermenting small, cooked soybeans with *Bacillus natto* until they develop a sticky, viscous coating. Natto is made from a number of specific varieties of soybeans and has a distinct taste and aroma that are an acquired taste for most. It is available only in Japanese food stores and is mostly imported, as very little is produced in the United States. It can be found frozen or fresh and will last about a week when refrigerated.

Soynuts

Soynuts are roasted soybeans that have been prepared by the dry or oil roasting of whole or split soybeans that have first been soaked in water. They can be sold with a coating of salt or other flavoring ingredients. Soynut pieces can be blended with other nuts and used in baking applications and other food preparations. Soynuts are high in protein, fiber, and the isoflavones found naturally in whole soybeans. Soynut butter is a paste of ground soynuts that is prepared in a manner similar to that of peanut butter and may have salt, sweeteners, and additional oil added. Soynuts and soynut butter can be found in many natural food stores and in some supermarkets.

Meat Alternatives

This is a broad product category, as literally hundreds of products are made from tofu, tempeh, textured soy flour, textured soy concentrate, isolated soy protein, and wheat gluten. Products take the shape of burgers, hot dogs, sausages, luncheon meats, chicken, fish, and ground meat. Most products are made with a combination of vegetable protein ingredients to achieve the best texture and are flavored for a particular use. Most are low in fat, and many are completely fat free. Meat alternatives are sold fresh and frozen and can be purchased in natural food stores, supermarkets, and, increasingly, in restaurants. Recent innovations include improved flavor and texture, as well as a new generation of extruded products that resemble muscle meat.

Cheese Alternatives

A wide range of cheese alternatives is made from soymilk, tofu, and other vegetable protein ingredients. These products can be found flavored to taste like American, mozzarella, cheddar, Monterey Jack, parmesan, and other cheeses. Most of these processed cheese products are made with a combination of soy and casein protein (from cow's milk). Casein is used because it is the protein responsible for the melting action common in cheese when it is heated. Without added casein, soy cheese alternatives would soften when heated but not melt or stretch. It also adds a flavor note that is associated with cheese. Some soy-based cheese alternatives have either had the fat replaced with vegetable oil or completely removed. These products are usually lactose and cholesterol free, as well. Soy cheese alternatives are also used as an ingredient in prepared frozen pizzas, stuffed in pasta, and added to frozen entrees.

Nondairy Frozen Desserts

This product category was one of the first to prove to Americans that products made from tofu and soybeans could taste good. Popularized by products such as Tofutti® brand (Tofutti Brands, Inc., Cranford, NJ) frozen desserts in the mid-1980s, this category has seen a recent reemergence due to the increased popularity of soymilk and premium dessert products. Nondairy frozen desserts are essentially ice cream products that are made using soymilk, tofu, or soy protein in place of dairy milk or cream. Both hard-pack and soft-serve styles exist, as well as novelty ice cream pops and sandwiches.

Green Vegetable Soybeans (Edamame)

This simple and nutritious soyfood is really just the whole soybean picked at its peak of maturity, at a time when it is high in sucrose and chlorophyll. It is harvested in the pod and is sold either in the pod or shucked, after being blanched and frozen. Because they are picked when their sugar levels are high, green vegetable soybeans are very sweet and pleasant tasting, and they have a firm texture. The common and traditional Japanese names for the green vegetable soybean are *edamame* when it is sold in the pod and *mukimame* when it is sold as individual beans. Green vegetable soybeans contain about 13% protein (the same amount as tofu) and are naturally high in calcium. They work very well in stir-fry dishes and can also be blended into dips and other preparations. They are becoming increasingly easy to find in the United States, as some American food processors are now packaging them for the domestic market.

Soy Sprouts

Soy sprouts are the fresh, crisp sprouts of germinated soybeans. They are a traditional food of Korea and may be eaten raw or in prepared food dishes. They are usually harvested after having grown for 5 to 7 days. They contain vitamin C and are high in protein and fiber. Soy sprouts are slightly larger than ordinary bean sprouts, which are prepared from mung bean seeds. Soy sprouts are available in some specialty food stores.

Full-Fat Soy Flour

Full-fat soy flour is made from whole or dehulled soybeans, which are ground into fine flour. Because it is made from the whole soybean, the composition of full-fat flour is identical to that of the natural soybean, with a protein content between 35 and 40% and a fat level between 15 and 20%. It is extremely nutritious and high in fiber and contains all of the vitamins, minerals, and phytochemicals of soybeans. Full-fat soy flour is available either as raw, enzyme-active flour or as toasted flour. The raw form is great for baking, as the active enzyme, lipoxygenase, helps to bleach wheat flour and produce a whiter bread loaf with improved moisture retention and shelf life. The toasted variety has a somewhat nuttier flavor and tends to have a slightly darker color. Enzyme-active full-fat soy flour can also be used to make fresh soymilk, which eliminates the need to soak or grind soybeans.

Defatted Soy Flour

Defatted soy flour, sometimes referred to simply as soy flour, is made by finely grinding defatted soybean flakes. The defatted flakes are produced by first crushing soybeans in a roller mill and then using a solvent to remove the oil. Any remaining solvent is then removed from the flake by heat and evaporation. Because the fat is thus removed, defatted soy flour has a higher protein content than full-fat soy flour. It contains 44 to 54% protein, 0.5 to 1.0% fat, 17 to 18% dietary fiber, and 30 to 35% total carbohydrates. (Defatted soy grits are processed from the same material but are not as finely ground.) Low-fat soy flours can also be produced using soybeans that have been mechanically pressed to remove the oil from soybeans. Because mechanical extraction of the oil is not as efficient as solvent extraction, these flours almost always contain some remaining fat, usually between 7 and 10%, and their protein content ranges from 45 to 50%. Defatted soy flour contains a higher level of isoflavones than either whole soybeans or products produced from them, due to the increased concentration of protein. Some of the nutrients attached to the oil — such as the vitamin E and lecithin — have been removed, but most of the vitamins, minerals, fiber, and phytochemicals have been retained.

Textured Soy Flour

Textured soy flour is probably the most popular form of defatted soy flour available to consumers. It is produced from defatted soy flour that is hydrated and then cooked under high pressure in an extruder to produce a variety of textured and shaped products. The extrusion process produces the unique texture by expanding the structure of the soy protein. A die at the end of the extruder determines the shape and size of the piece or granule. Flavor or color can be added to the soy flour prior to processing. Available in a range of colors, shapes, and sizes, textured soy flour, or textured soy protein (also known as TVP®, a registered trademark of the Archer Daniels Midland Company), is most commonly used as a meat extender or replacement due to its meat-like appearance and consistency when prepared for use. To use textured soy flour, one must first hydrate the granules with hot or boiling water. As the textured soy flour absorbs the water, it expands and softens.

Soy Protein Concentrate

Soy protein concentrate is comprised of between 65 and 70% soy protein, with trace amounts of fat and 5 to 6% fiber. It is made by processing defatted soybean flakes in either an alcohol or water bath to remove soluble sugars. The result is a concentrated form of soy flour with improved flavor and functional characteristics. It can be texturized (the most common form found in meat alternatives) or spray dried to form a powder used in infant formulas and other dairy alternative applications. Due to the alcohol-washing step used to reduce the sugars, most of the isoflavones are removed during processing, although the quality of the protein is not reduced. It is also more easily digested than soy flour as most of the sugars responsible for creating flatulence are removed in the processing.

Soy Protein Isolate

Soy protein isolate, or isolated soy protein, is 85 to 90% pure soybean protein, processed from the same defatted soy flakes as defatted soy flour and soy protein concentrate, but isolated soy proteins are processed one step beyond concentrates to remove not only the fat and soluble sugars but also the insoluble sugars and dietary fiber. Isolated soy proteins are very low in flavor, highly digestible, and easy to use in food, beverage, and baking formulations. They disperse easily in water and work well as emulsifiers, helping to bind water and fat together. Isolated soy proteins are found in many of the meat and dairy alternatives on the market today. They are also used in conjunction with whole soybeans to make certain types of tofu, are blended with tofu in tofu hot dogs to improve the texture, are used in some soymilk products in place of the whole soybean, and are used in nondairy creamers and infant formulas. They are also found in many of the weight- and muscle-building products sold to the fitness market. Due to their highly functional qualities, the meat industry uses isolated soy proteins to help bind water and fat in processed meats and to reduce shrinkage during cooking. Consumers can buy isolated soy proteins under a variety of names and labels, marketed as a nutritional aid or a base ingredient for food and beverages. They are most likely to be found in natural food or supplement stores and can be purchased through some direct mail catalogs. Properly stored, they should have a long shelf life with or without refrigeration.

Conclusion

One would be hard pressed to find a more valuable resource for our planet's food supply than the humble soybean. With its healthfulness proven by its widespread use for thousands of years and its ample supply and low cost, the soybean is proving itself to be a food for all humankind. For the next few decades, it is likely that we will see continual improvements in plant genetics and processing technology that will make the soybean an even more valuable resource for food, feed, and a myriad of innovative industrial uses as well. Given its long history and tremendous versatility, the soybean is indeed proving itself to be a new food platform for the 21st century.

2

Overview of the Health Effects of Soyfoods

Mark Messina

CONTENTS

Introduction

Foods made from the soybean have been consumed for centuries in Asia and by vegetarians in non-Asian countries for decades. The soybean has long been embraced as a source of high-quality protein from which a wide variety of foods can be made. The quality of soy protein is now recognized as being essentially equivalent to that of animal protein, and for this reason, in the year 2000, the U.S. Department of Agriculture (USDA) issued a ruling allowing soy protein to completely replace animal protein in the National School Lunch Program; previously, soy protein could replace no more than 30% of animal protein.[1]

During the past 15 years, the popularity of both traditional and Western-style soy products has increased dramatically in the United States. This popularity has little to do with the protein quality of the soybean; rather, it is because of research suggesting that the consumption of fairly modest amounts of soyfoods is associated with health benefits in a variety of areas. In fact, in 1999, the U.S. Food and Drug Administration approved a health claim for the cholesterol-lowering effects of soy protein.[2] Beyond heart disease, proposed benefits include reductions in the risk of breast and prostate cancer[3] and osteoporosis.[4] More speculative data suggest soyfoods may also favorably affect kidney function[5] and cognitive function[6] and help to alleviate hot flashes in menopausal women.[7]

In addition to an assortment of vitamins and minerals, like all plant foods the soybean contains numerous biologically active nonnutritive substances.[8] Certainly, it is reasonable to think that collectively these components (nutrients and nonnutrients) account for the hypothesized health benefits of soyfoods; however, much evidence suggests that one particular class of compounds — isoflavones — is primarily responsible for many of these benefits, although the protein is required for cholesterol reduction.

Isoflavones are classified as phytoestrogens because they bind to estrogen receptors,[9] but they are very complex molecules that not only differ from estrogen but also have potentially important nonhormonal properties as well.[10] Furthermore, in comparison to estrogen receptor alpha, isoflavones preferentially bind to[9] and activate estrogen receptor beta (ERβ).[11,12] The preferential binding of isoflavones to ERβ likely contributes to evidence suggesting that isoflavones function as selective estrogen receptor modulators.[13]

The "estrogen-like" qualities of isoflavones have caused them to be viewed as possible alternatives to conventional hormonal therapy, and for this reason many women experiencing menopause have been attracted to soyfoods. Isoflavones are present in many different plants, but the soybean is the only commonly consumed food to contain nutritionally relevant amounts of these diphenolic molecules. Consumer interest in isoflavones has led to the development of isoflavone supplements and the use of isoflavones as food fortificants.

The purpose of this review is to evaluate the strength of the evidence for the different proposed benefits of soy and to make intake recommendations on the basis of this information. Due to space limitations only cancer, osteoporosis, coronary heart disease, and hot flashes are discussed.

Soybean Isoflavone Content and Profile

Isoflavones are a subclass of flavonoids that have a very limited distribution in nature and are found in few commonly consumed foods, which accounts for surveys showing that among typical non-Asians isoflavone intake is

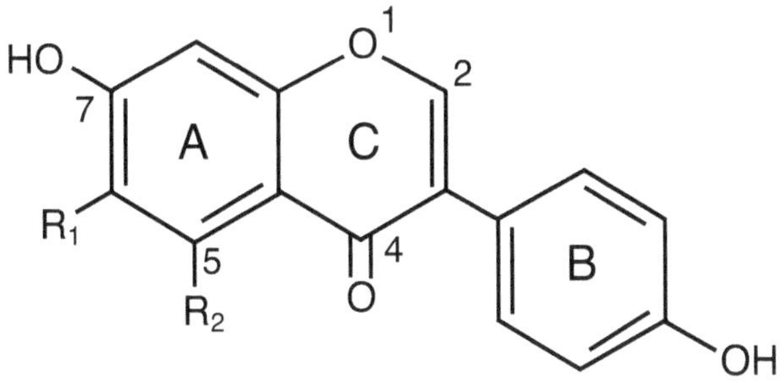

	R$_1$	R$_2$
Genistein	H	OH
Daidzein	H	H
Glycitein	OCH$_3$	H

FIGURE 2.1
Structures of the three aglycone soybean isoflavones.

negligible. The 15-carbon (C6–C3–C6) backbone of flavonoids can be arranged as a 1,3-diphenylpropane skeleton (flavonoids nucleus) or as a 1,2-diphenyl-propane nucleus skeleton (isoflavonoid nucleus). In isoflavones, the B-ring is attached to the C-ring at the 3- rather than the 2-position (Figure 2.1). The three soybean isoflavones are genistein (4′,5,7-trihydroxyisoflavone), daidzein (4′,7-dihydroxyisoflavone), and glycitein (7,4′-dihydroxy-6-methoxyisofla-vone). Typically, more genistein exists in soybeans and soyfoods than daid-zein, whereas glycitein comprises less than 10% of the total isoflavone content of the soybean.[14] Isoflavone supplements made from the hypocotyledon (often referred to as soy germ) portion of the soybean are rich in glycitein and low in genistein. In contrast, isoflavone supplements derived from soy molasses have an isoflavone profile similar to that found in the soybean.

Soybeans contain approximately 1.2 to 3.3 mg isoflavones per g dry weight, and every gram of protein in traditional soyfoods is associated with approximately 3.5 mg isoflavones; however, processing can dramatically reduce isoflavone content, especially in the case of alcohol-extracted soy protein concentrate and isolate (Table 2.1). Consequently, it is difficult to predict the isoflavone content of soy protein without knowing the process used to the make the specific product. For example, isolated soy proteins have an isoflavone content that ranges from 0.5 to 2.0 mg/g. In conjunction with Iowa State University, the USDA has created an online database of the isoflavone content of foods (http://www.nal.usda.gov/fnic/foodcomp/Data/isoflav/isoflav.html).

Isoflavones are naturally present in the soybean primarily in their beta glycoside form (genistein, daidzein, and glycitein). The sugar molecule is attached at the 7-position of the A ring. An acetyl or malonyl group can be

TABLE 2.1

Isoflavone (Aglycone Weight) Content of Selected Foods

Food (Nutrient Database Number)	Samples	Isoflavone Content (mg/100 g edible portion)		
		Mean	Minimum	Maximum
Tofu				
Firm (16126)	6	24.7	7.9	34.6
Regular (16427)	4	23.6	5.1	33.7
Silken, firm (16162)	2	27.9	23.8	32.0
Natto (16113)	5	58.9	46.4	87.0
Soymilk (16120)	14	9.7	1.3	21.1
Miso (16112)	7	42.6	22.70	89.2
Tempeh (16114)	6	43.5	6.9	62.5
Soynuts, dry roasted (16111)	7	128.4	1.66	201.9
Soybeans, cooked (16109)	1	54.7	NA	NA
Isolated soy protein (16122)	14	97.43	46.5	199.3
Soy protein concentrate (aqueous washed) (99060)	3	102.1	61.2	167.0
Soy protein concentrate (alcohol washed) (16121)	5	12.5	2.1	31.8

Source: U.S. Department of Agriculture, Washington, D.C. (http://www.nal.usda.gov/fnic/foodcomp/index.html).

attached to the glucose molecule, resulting in a total of 12 different soybean isoflavones isoforms. Heating does not cause a significant loss of isoflavones, although it can result in decarboxylation, thereby converting the malonyl isoflavone glycosides into acetyl glycosides. Perhaps more relevant from a physiological perspective because of the possible impact on bioavailability (see next section) is the conversion (primarily as a result of bacterial hydrolysis during fermentation) of the glycosides into aglycone isoflavones.

Finally, because the weight of the aglycone isoflavone, the biologically active isomer, is only approximately 60% that of the glycoside, unless specifically indicated 100 mg isoflavones can refer to between 60 and 100 mg of active molecule. To avoid confusion, the recommendation has been made to use the aglycone weight.[15] This is the case in the text that follows.

Isoflavone Absorption and Metabolism

Isoflavone glycosides are not absorbed intact, but hydrolysis (from the acid pH of the stomach, endogenous enzymes, and intestinal microflora) does readily occur *in vivo*; thus, there is little difference in bioavailability between the glycoside and aglycone forms of isoflavones.[16] After the ingestion of soyfoods, a small peak in serum levels occurs approximately 1 to 2 hours later, but the major serum peak occurs 4 to 6 hours after ingestion. Most work estimates that the half-life of isoflavones is between 4 and 8 hours; 24 hours after the consumption of soyfoods nearly all of the isoflavones are excreted.[16] Serum isoflavone levels increase in a dose-dependent fashion in response to soyfood consumption.[17,18] The highest sustained serum isoflavone levels are likely achieved by dividing daily isoflavone intake into several doses. This method of ingestion reflects the pattern of isoflavone intake from soyfoods in Asia. A huge variation in isoflavone metabolism exists among individuals. This variation leads to very different serum levels of both isoflavones and their metabolites in response to the ingestion of isoflavones.[19] This variation likely contributes to the inconsistency noted in the literature regarding the health effects of isoflavones. An especially potentially important observation is that only approximately 30 to 50% of subjects possess the intestinal bacteria that convert the isoflavone daidzein into the isoflavonoid equol, which has been proposed as an especially beneficial compound.[20]

Asian Soy Protein and Isoflavone Intake

Part of the enthusiasm for the health benefits of soy, rightly or wrongly, is based on the low rates of breast and prostate cancer[21] and coronary heart disease (CHD)[22] in Asia, particularly Japan. Thus, arguably, one basis for determining Western soy intake recommendations is Asian soy consumption. Although widely varying estimates of Asian soy intake have been reported in the literature within the past 7 years, many large surveys of soy protein and isoflavone consumption by Asian adults have been published. These surveys, which often include as many as nine different questions related to soy, provide a very accurate picture of actual intake in Asia. It is clear from these data that early soy intake estimates were greatly exaggerated. Surveys suggest that older (50 years) Japanese adults typically consume 7 to 11 g soy protein and 30 to 50 mg isoflavones per day. Intake in Hong Kong and Singapore is lower than in Japan, and significant regional intake differences exist for China. Evidence suggests ≤10% of the Asian population consumes as much as 25 g soy protein or 100 mg isoflavones per day.

Soy and the Risk of Chronic Diseases

Cancer

The National Cancer Institute in the United States has been actively investigating the anticancer effects of soy since 1991, when the first request for applications on this subject was issued.[8] In part, this interest in soy stems from the historically low incidence rates of breast and prostate cancer in Asia.[21] Evidence suggests that the potential anticancer effects of soy extend beyond these two cancers, but, because most research has focused only on breast and prostate cancer, these are the cancers discussed in this chapter. Several putative anticarcinogens can be found in soybeans and soyfoods;[8] however, the soybean isoflavones have received the most attention.[10] With regard to both cancer prevention and treatment, the effects of genistein on signal transduction are of particular interest.[10] Genistein inhibits the growth of a wide range of cancer cells *in vitro* by inhibiting the activity of certain enzymes, inducing apoptosis, and inhibiting the activation of NF-κB and Akt signaling pathways;[10] however, the possible antiestrogenic effects of isoflavones — which were first demonstrated in rodents almost 40 years ago — provide an additional possible mechanism for the hypothesized anticancer effects of soy.[23]

Breast Cancer

The relationship between soy intake and breast cancer risk has been investigated extensively. Animal studies generally suggest that soy is protective against this cancer; however, the epidemiologic data are only weakly or not at all supportive of this hypothesis.[24] Furthermore, clinical studies investigating markers of breast cancer risk, such as breast tissue density and serum estrogen levels, have also produced relatively unimpressive results.[25,26] Thus, overall, it is difficult to conclude that adult soy intake reduces breast cancer risk; however, provocative data suggest that intake during adolescence reduces breast cancer risk later in life.

Two research groups from the United States — Larmatiniere and colleagues from the University of Alabama and Hilakivi-Clarke and colleagues from Georgetown University in Washington, D.C. — are responsible for all of the animal work examining the effects of early genistein exposure on mammary carcinogenesis. Both teams conducted their research in Sprague–Dawley rats, and each used the indirect acting carcinogen 7,12-dimethylbenz(a)anthracene to initiate mammary tumors. These studies show that exposing rats to genistein during the first three weeks of life reduces tumor multiplicity by approximately one half (Table 2.2).[27,28] Interestingly, Lamartiniere et al. have shown that, although genistein administration during adulthood has little impact on tumor development, when it is given to rats also exposed to genistein when they were young, tumor number is suppressed significantly beyond that achieved with early intake alone (Table 2.2).

Only two epidemiologic studies have examined the "early soy" hypothesis, but both studies are consistent with the animal data. A large (1459 cases and 1556 controls) Chinese case-control study found that soy protein consumption during adolescence reduced adult breast cancer risk by approximately 50%, but adult soy consumption had no impact.[29] In this study, the intake cutoff of the fourth quartile was only 11.19 g/d soy protein. Also in

TABLE 2.2

Effect of Genistein on Development of
7,12-Dimethylbenz(a)anthracene-
Induced Mammary Tumors in Rats

Exposure Period	Tumors/Rat
None	8.9
Prenatal (*in utero*)	8.8
Adult (PND 100–180)	8.2
Prepubertal (PND 1–21)	4.3
Prepubertal and adult	2.8

Note: Nursing dams and adults were fed
diets containing 250 mg genistein per
kg. PND, postnatal day.
Source: Lamartiniere, C.A. et al., *J. Women's
Cancer*, 2, 11–19, 2000. With permission.

support of this hypothesis is a U.S. study (501 cases and 594 controls) involving Asian– Americans that found high soy consumption throughout life was associated with a one third reduction in risk of breast cancer, whereas high adult intake alone was not protective.[30]

Prostate Cancer

In 2000, the International Prostate Health Council suggested that isoflavones prevent latent prostate cancer from progressing to the more advanced form of this disease.[31] Recent animal work supports this conclusion.[32] Overall, with few exceptions, animal studies show that isoflavones and isoflavone-rich soy protein inhibit prostate tumors induced by chemical carcinogens or via the implantation of prostate cancer cells.[3] Interestingly, isoflavones in combination with tea extracts were shown to reduce tumor growth in mice implanted with androgen-sensitive prostate cancer cells more effectively than either agent alone.[33] Both soyfoods and tea are important components of the Asian diet.

Very limited epidemiologic investigation of the relationship between soy intake and prostate cancer risk has been conducted, although the data are generally supportive of the hypothesis that soy is protective. A recent analysis of ten epidemiologic studies found that soy intake was associated with a one third reduction in prostate cancer risk.[34] However, the limitations of the epidemiologic data in terms of both quantity and quality should not be overlooked. Many of the epidemiologic studies did not comprehensively evaluate soyfood intake, although a recent Japanese case-control study that did found that, when comparing the highest with the lowest soyfood intake, quartile risk was reduced by nearly 50%.[35]

The mechanism by which soy may reduce prostate cancer risk has not been identified, but soy does not appear to lower serum testosterone levels,[36,37] although in mice equol was recently shown to bind to and inactivate

dihydrotestosterone, the active metabolite of testosterone within the prostate.[38] Finally, in several studies neither soy nor isoflavones were found to
lower prostate-specific antigen (PSA) levels — a marker of prostate cancer
risk — in healthy subjects,[39,40] although several very preliminary studies have
found that soy or isoflavones may favorably affect PSA levels in prostate
cancer patients.[41,42]

Osteoporosis

The first human study suggesting that isoflavones favorably affect bone
health was published in 1998.[43] Since then, at least 14 additional publications
have examined the effects of soyfoods, isoflavone-rich soy protein, or isolated
isoflavones on bone loss in perimenopausal (N-1) or postmenopausal (N-14)
women.[4] As reviewed by Messina et al.,[4] these trials were conducted in 9
countries, included from 10 to 75 subjects per group (although most involved
more than 30), and, with one exception, were conducted for less than 1 year.
The results from these studies are mixed but overall indicate that isoflavones
reduce bone loss; however, the relatively small size and the short duration
of these osteoporosis trials prevent drawing definitive conclusions. To best
predict likely long-term effects, trials should be conducted for 2 to 3 years.[44]
The disappointing results from a 3-year trial involving the synthetic isoflavone ipriflavone are certainly evidence of this.[45] Nevertheless, the results from
these isoflavone trials are sufficiently encouraging to recommend that women
concerned about their bone health consider using isoflavones to help ward
off osteoporosis although they should not be used as a substitute for established antiosteoporotic medications. This conclusion is consistent with the
Asian epidemiologic data, which generally show that among Asian women
higher soy/isoflavone intake is associated with higher bone mineral density.[4]
Little dose–response information is available, but most trials examining the
skeletal system have used approximately 80 mg total isoflavones per day. The
means by which bone loss is reduced has not been identified, but evidence
suggests both hormonal and nonhormonal mechanisms. Large, long-term
trials of isoflavones and bone health are underway.

Coronary Heart Disease

The cholesterol-lowering effects of soy protein were first demonstrated in
humans in 1967.[46] Ten years later, a landmark paper by Italian researchers
on the hypocholesterolemic effects of soy protein was published in the journal *Lancet*.[47] Despite these reports, however, and others published over the
next 15 years, it was not until 1995 that soy received widespread attention.[48]
That year, Anderson and colleagues published a metaanalysis in the *New
England Journal of Medicine* which summarized the existing data on soy and
cholesterol.[48] They found that soy protein lowered cholesterol in 34 of the
38 trials and that the average decrease in low-density lipoprotein cholesterol
(LDLC) levels was an impressive 12.9%. The metaanalysis also found soy

protein modestly raised high-density lipoprotein cholesterol, a finding that has been confirmed by other research.[49]

The metaanalysis comprised the bulk of the research on which the FDA based its approval for a health claim for soy protein 4 years later (the 14 clinical studies with the best experimental designs were given the highest priority[50]), although it is now generally recognized that the initial estimates of the potency of soy protein were exaggerated. In 2003, a metaanalysis by Weggemans and Trautwein[51] that included 10 clinical trials and involved nearly 1000 subjects published after 1995 found that the average reduction in LDLC in response to soy protein was only 4%. These more modest effects are still clinically relevant, however. Some estimates suggest that each 1% reduction in cholesterol reduces CHD risk by as much as 3 to 4%.[52,53] Furthermore, soy protein can be combined with other dietary approaches to effectively lower cholesterol.[54] Also, the use of soy protein along with a statin might help to avoid doubling the dose of medication when the target cholesterol goal is not achieved with the medication alone, thereby reducing the possibility of adverse drug effects.

When the FDA approved the soy protein health claim, they established 25 g/day as the threshold intake level necessary for cholesterol reduction. The consumption of this amount of soy protein represents a significant dietary challenge for non-Asians, as this figure is approximately 2.5 times the typical Japanese intake. The FDA set 25 g/day as the required intake not because evidence suggested that lower amounts were not efficacious but because few trials used amounts lower than this. In fact, provocative evidence indicates that fewer than 25 g/day is hypocholesterolemic.[55]

A far more contentious issue is the extent, if any, to which isoflavones impact the cholesterol-lowering effects of soy protein (isoflavones by themselves do not lower cholesterol). The position of the FDA is that the evidence does not warrant concluding that isoflavones play a role in cholesterol reduction, although a recent metaanalysis suggests otherwise.[56] A leading hypothesis for the hypocholesterolemic effects of soy is that the peptides resulting from the ingestion of soy protein upregulate hepatic LDLC receptors.[57,58]

Importantly, the role of soy in reducing CHD risk may extend far beyond cholesterol reduction. In fact, it may be that the possible coronary benefits of isoflavones exceed those of soy protein, although this remains speculative. Several studies have found that isoflavones enhance endothelial function[59] and systemic arterial compliance,[60] both of which are considered to be indicators of coronary health.[61,62] In addition, isoflavones and their metabolites are antioxidants,[63] and speculative data suggest that isoflavone-rich soy protein inhibits LDL-oxidation[64,65] and perhaps platelet aggregation.[66] Indirect support for the coronary benefits of isoflavones comes from several Asian epidemiologic studies that found soyfood intake to be strongly inversely related to the risk of coronary events; the reduction in risk was far beyond that which could be attributable to cholesterol reduction alone.[67–70] It is clear that the many possible coronary benefits warrant adding soyfoods to a heart-healthy diet.

Hot Flashes

For the majority of women, hot flashes begin prior to menopause, and about 10 to 15% of women who have hot flashes have them very frequently and severely.[71] Both clinical[7] and epidemiologic[72] data suggest that isoflavones help to alleviate hot flashes. The percentage of women in Asian countries, especially Japan, who experience hot flashes is quite low relative to the West.[73] In 2003, Messina and Hughes[7] published a review of 19 trials that involved over 1700 women and examined the effects of soyfoods or isoflavone supplements on menopausal symptoms. After excluding six trials from their regression analysis for methodological reasons, they found among the remaining 13 trials a statistically significant relationship between initial hot flash frequency and treatment efficacy. According to these findings much of the inconsistency in results from the clinical trials can be attributed to the variation in mean initial hot flash frequency among the studies. No dose–response information from the trials in this analysis is available, but the isoflavone doses used in the clinical trials range from 34 to 100 mg/day. In contrast to this analysis, however, Krebs et al.[74] found no support for the ability of isoflavones (from both red clover and soybeans) to alleviate hot flashes. Their analysis included 25 trials involving 2348 participants who had a mean daily hot flash frequency of 7.1. Thus, at this point it is not possible to conclude that isoflavones are helpful for alleviating hot flashes, but, arguably, the data are sufficiently suggestive for health professionals to suggest that women try isoflavones for relief. This recommendation is partially justified because of the possible coronary and skeletal benefits of isoflavones, as discussed previously.

Intake Recommendations for Healthy Adults

The only soy intake recommendation from an established health organization comes from the FDA, which recommends a soy protein intake of 25 g/day for cholesterol reduction. This recommendation should not form the basis for soy intake among the general population, however, because it pertains to only one health attribute of soy and is unrelated to isoflavone intake.[50] If soy contributes to the lower breast and prostate cancer incidence in Asia, then the average daily Japanese intakes of 30 to 50 mg isoflavones may be efficacious, but the mean intake may significantly underestimate the amount needed for maximum protection. The epidemiologic studies demonstrating reductions in the risk of coronary heart disease and cancer involved comparisons across intake categories, and the largest reductions in risk are typically associated with intakes greater than the mean. Thus, maximum protection might require the consumption of closer to 75 mg/day of isoflavones. This higher value is similar to the level of isoflavones used in many of the bone trials in postmenopausal women and is consistent with the amounts used in many of the hot flash trials showing isoflavones to be efficacious.

Safety Concerns

As more information about the pleotrophic effects of isoflavones has become available, concerns about the possible adverse effects of soyfood consumption have been raised. Much of this concern is focused on the effects of isoflavone exposure during development as a result of *in utero* exposure and exposure during infancy via soy infant formula.[75] These research areas are beyond the scope of this chapter; however, concerns have also been raised about the effect of soy on thyroid function and the effect of soy in breast cancer patients and women at high risk of this disease. The breast cancer concern is based on the estrogen-like effects of isoflavones and data showing that isolated soy protein and isoflavones stimulate tumor growth in ovariectomized athymic mice implanted with estrogen-positive breast cancer cells;[76,77] however, data in humans generally do not support these concerns.[26,78,79] Furthermore, evidence suggests that the ovariectomized athymic mouse model represents a hypoestrogenic environment that does not reflect conditions in either pre- or postmenopausal women.[80] Nevertheless, this issue remains controversial and breast cancer patients are advised to discuss this matter with the appropriate healthcare provider. The relationship between soy consumption and thyroid function has been investigated for 70 years. Important insight into this relationship has come from the numerous clinical trials conducted during the past several years.[81] These data suggest neither soy nor isoflavones affect thyroid function in healthy adults; however, preliminary data suggest that soyfood consumption may increase the dose of thyroid medication required by hypothyroid patients.[82] Even in these patients, however, soyfoods need not be avoided.

Conclusions

Firm conclusions about the ability of soy to reduce the risk of chronic disease cannot be made. This is not surprising, as definitive data can come only from long-term randomized clinical trials that include disease events as endpoints. Unfortunately, the cost and complexity of such trials are enormous; consequently, few such trials can be conducted. To date, information about the health effects of soy comes from animal and epidemiologic studies and from intervention trials in which markers of disease risk have been evaluated.

Nevertheless, the evidence in several different areas is sufficiently encouraging for health professionals to encourage greater soy consumption among Western populations. Based on Asian intake and the results from epidemiologic and intervention trials, a reasonable intake recommendation is 15 to 20 g soy protein and 50 to 75 mg isoflavones per day. This amount of soy is provided by approximately 2 to 3 servings of traditional Asian soyfoods — a significant dietary challenge for non-Asians. However, with the development of new soy products that mimic the kinds of foods common to Western diets, progress toward this goal can be made.

References

1. USDA, Modification of the vegetable protein products requirements for the National School Lunch program, School Breakfast program, Summer Food Service program and Child and Adult Care Food program, CFR Parts 210, 215, 220, 225, 226, *Fed. Reg.*, 7, 12429–12442, 2000.
2. Food labeling: health claims. soy protein and coronary heart disease, *Fed. Reg.*, 64(206), 57699–57733, 1999.
3. Messina, M.J., Emerging evidence on the role of soy in reducing prostate cancer risk, *Nutr. Rev.*, 61, 117–131, 2003.
4. Messina, M.J., Ho, S., and Alekel, D.L., Skeletal benefits of soy isoflavones: a review of the clinical trial and epidemiologic data, *Curr. Opin. Clin. Nutr. Metab. Care*, 7, 649–658, 2004.
5. Velasquez, M.T. and Bhathena, S.J., Dietary phytoestrogens: a possible role in renal disease protection, *Am. J. Kidney Dis.*, 37, 1056–1068, 2001.
6. Duffy, R., Wiseman, H., and File, S.E., Improved cognitive function in postmenopausal women after 12 weeks of consumption of a soya extract containing isoflavones, *Pharmacol. Biochem. Behav.*, 75, 721–729, 2003.
7. Messina, M.J. and Hughes, C., Efficacy of soyfoods and soybean isoflavone supplements for alleviating menopausal symptoms is positively related to initial hot flush frequency, *J. Med. Food*, 6, 1–11, 2003.
8. Messina, M.J. and Barnes, S., The role of soy products in reducing risk of cancer, *J. Natl. Cancer Inst.*, 83, 541–546, 1991.
9. Kuiper, G.G., Lemmen, J.G., Carlsson, B. et al., Interaction of estrogenic chemicals and phytoestrogens with estrogen receptor beta, *Endocrinology*, 139, 4252–4263, 1998.
10. Sarkar, F.H. and Li, Y., Soy isoflavones and cancer prevention, *Cancer Invest.*, 21, 744–757, 2003.
11. An, J., Tzagarakis-Foster, C., Scharschmidt, T.C., Lomri, N., and Leitman, D.C., Estrogen receptor beta: selective transcriptional activity and recruitment of coregulators by phytoestrogens, *J. Biol. Chem.*, 276, 17808–17814, 2001.
12. Kostelac, D., Rechkemmer, G., and Briviba, K., Phytoestrogens modulate binding response of estrogen receptors alpha and beta to the estrogen response element, *J. Agric. Food Chem.*, 51, 7632–7635, 2003.

13. Diel, P., Geis, R.B., Caldarelli, A. et al., The differential ability of the phytoestrogen genistein and of estradiol to induce uterine weight and proliferation in the rat is associated with a substance specific modulation of uterine gene expression, *Mol. Cell. Endocrinol.*, 221, 21–32, 2004.

14. Murphy, P.A., Song, T., Buseman, G. et al., Isoflavones in retail and institutional soy foods, *J. Agric. Food Chem.*, 47, 2697–2704, 1999.

15. Erdman, Jr., J.W., Badger, T.M., Lampe, J.W., Setchell, K.D., and Messina, M.J., Not all soy products are created equal: caution needed in interpretation of research results, *J. Nutr.*, 134, 1229S–1233S, 2004.

16. Rowland, I., Faughnan, M., Hoey, L., Wahala, K., Williamson, G., and Cassidy, A., Bioavailability of phyto-oestrogens, *Br. J. Nutr.*, 89(Suppl. 1), S45–S58, 2003.

17. Xu, X., Harris, K.S., Wang, H.J., Murphy, P.A., and Hendrich, S., Bioavailability of soybean isoflavones depends upon gut microflora in women, *J. Nutr.*, 125, 2307–2315, 1995.

18. Xu, X., Wang, H.J., Murphy, P.A., Cook, L., and Hendrich, S., Daidzein is a more bioavailable soymilk isoflavone than is genistein in adult women, *J. Nutr.*, 124, 825–832, 1994.

19. Wiseman, H., Casey, K., Bowey, E.A. et al., Influence of 10 weeks of soy consumption on plasma concentrations and excretion of isoflavonoids and on gut microflora metabolism in healthy adults, *Am. J. Clin. Nutr.*, 80, 692–699, 2004.

20. Setchell, K.D., Brown, N.M., and Lydeking-Olsen, E., The clinical importance of the metabolite equol: a clue to the effectiveness of soy and its isoflavones, *J. Nutr.*, 132, 3577–3584, 2002.

21. Pisani, P., Bray, F., and Parkin, D.M., Estimates of the world-wide prevalence of cancer for 25 sites in the adult population, *Int. J. Cancer*, 97, 72–81, 2002.

22. Saito, I., Folsom, A.R., Aono, H., Ozawa, H., Ikebe, T., and Yamashita, T., Comparison of fatal coronary heart disease occurrence based on population surveys in Japan and the USA, *Int. J. Epidemiol.*, 29, 837–844, 2000.

23. Folman, Y. and Pope, G.S., The interaction in the immature mouse of potent oestrogens with coumestrol, genistein and other utero-vaginotrophic compounds of low potency, *J. Endocrinol.*, 34, 215–225, 1966.

24. Messina, M.J. and Loprinzi, C.L., Soy for breast cancer survivors: a critical review of the literature, *J. Nutr.*, 131, 3095S–108S, 2001.

25. Brown, B.D., Thomas, W., Hutchins, A., Martini, M.C., and Slavin, J.L., Types of dietary fat and soy minimally affect hormones and biomarkers associated with breast cancer risk in premenopausal women, *Nutr. Cancer*, 43, 22–30, 2002.

26. Maskarinec, G., Takata, Y., Franke, A.A., Williams, A.E., and Murphy, S.P., A 2-year soy intervention in premenopausal women does not change mammographic densities, *J. Nutr.*, 134, 3089–3094, 2004.

27. Lamartiniere, C.A., Zhao, Y.X., and Fritz, W.A., Genistein: mammary cancer chemoprevention, *in vivo* mechanisms of action, potential for toxicity and bioavailability in rats, *J. Women's Cancer*, 2, 11–19, 2000.

28. Hilakivi-Clarke, L., Onojafe, I., Raygada, M. et al., Prepubertal exposure to zearalenone or genistein reduces mammary tumorigenesis, *Br. J. Cancer*, 80, 1682–1688, 1999.

29. Shu, X.O., Jin, F., Dai, Q. et al., Soyfood intake during adolescence and subsequent risk of breast cancer among Chinese women, *Cancer Epidemiol. Biomarkers Prev.*, 10, 483–488, 2001.

30. Wu, A.H., Wan, P., Hankin, J., Tseng, C.C., Yu, M.C., and Pike, M.C., Adolescent and adult soy intake and risk of breast cancer in Asian-Americans, *Carcinogenesis*, 23, 1491–1496, 2002.

31. Griffiths, K., Estrogens and prostatic disease, International Prostate Health Council Study Group, *Prostate*, 45, 87–100, 2000.

32. Hikosaka, A., Asamoto, M., Hokaiwado, N. et al., Inhibitory effects of soy isoflavones on rat prostate carcinogenesis induced by 2-amino-1-methyl-6-phenylimidazo[4,5-b]pyridine (PhIP), *Carcinogenesis*, 25, 381–387, 2004.

33. Zhou, J.R., Yu, L., Mai, Z., and Blackburn, G.L., Combined inhibition of estrogen-dependent human breast carcinoma by soy and tea bioactive components in mice, *Int. J. Cancer*, 108, 8–14, 2004.

34. *The Health Claim Petition: Soy Protein and the Reduced Risk of Certain Cancers*, The Solae Company, St. Louis, MO, 2004.

35. Lee, M.M., Gomez, S.L., Chang, J.S., Wey, M., Wang, R.T., and Hsing, A.W., Soy and isoflavone consumption in relation to prostate cancer risk in China, *Cancer Epidemiol. Biomarkers Prev.*, 12, 665–668, 2003.

36. Lewis, J.G., Morris, J.C., Clark, B.M., and Elder, P.A., The effect of isoflavone extract ingestion, as Trinovin, on plasma steroids in normal men, *Steroids*, 67, 25–29, 2002.

37. Mitchell, J.H., Cawood, E., Kinniburgh, D., Provan, A., Collins, A.R., and Irvine, D.S., Effect of a phytoestrogen food supplement on reproductive health in normal males, *Clin. Sci. (Lond.)*, 100, 613–618, 2001.

38. Lund, T.D., Munson, D.J., Haldy, M.E., Setchell, K.D., Lephart, E.D., and Handa, R.J., Equol is a novel anti-androgen that inhibits prostate growth and hormone feedback, *Biol. Reprod.*, 70, 1188–1195, 2004.

39. Urban, D., Irwin, W., Kirk, M. et al., The effect of isolated soy protein on plasma biomarkers in elderly men with elevated serum prostate specific antigen, *J. Urol.*, 165, 294–300, 2001.

40. Adams, K.F., Chen, C., Newton, K.M., Potter, J.D., and Lampe, J.W., Soy isoflavones do not modulate prostate-specific antigen concentrations in older men in a randomized controlled trial, *Cancer Epidemiol. Biomarkers Prev.*, 13, 644–648, 2004.

41. Hussain, M., Banerjee, M., Sarkar, F.H. et al., Soy isoflavones in the treatment of prostate cancer, *Nutr. Cancer*, 47, 111–117, 2003.

42. Dalais, F.S., Meliala, A., Wattanapenpaiboon, N. et al., Effects of a diet rich in phytoestrogens on prostate-specific antigen and sex hormones in men diagnosed with prostate cancer, *Urology*, 64, 510–515, 2004.

43. Potter, S.M., Baum, J.A., Teng, H., Stillman, R.J., Shay, N.F., and Erdman, Jr., J.W., Soy protein and isoflavones: their effects on blood lipids and bone density in postmenopausal women, *Am. J. Clin. Nutr.*, 68, 1375S–1379S, 1998.

44. Heaney, R.P., Calcium, dairy products and osteoporosis, *J. Am. Coll. Nutr.*, 19, 83S–99S, 2000.

45. Alexandersen, P., Toussaint, A., Christiansen, C. et al., Ipriflavone in the treatment of postmenopausal osteoporosis: a randomized controlled trial, *JAMA*, 285, 1482–1488, 2001.

46. Hodges, R.E., Krehl, W.A., Stone, D.B., and Lopez, A., Dietary carbohydrates and low cholesterol diets: effects on serum lipids on man, *Am. J. Clin. Nutr.*, 20, 198–208, 1967.

47. Sirtori, C.R., Agradi, E., Conti, F., Mantero, O., and Gatti, E., Soybean-protein diet in the treatment of type-II hyperlipoproteinaemia, *Lancet*, 1, 275–277, 1977.

48. Anderson, J.W., Johnstone, B.M., and Cook-Newell, M.E., Meta-analysis of the effects of soy protein intake on serum lipids, *N. Engl. J. Med.*, 333, 276–282, 1995.
49. Greany, K.A., Nettleton, J.A., Wangen, K.E., Thomas, W., and Kurzer, M.S., Probiotic consumption does not enhance the cholesterol-lowering effect of soy in postmenopausal women, *J. Nutr.*, 134, 3277–3283, 2004.
50. FDA, Food labeling, health claims, soy protein, and coronary heart disease, *Fed. Reg.*, 57, 699–733, 1999.
51. Weggemans, R.M. and Trautwein, E.A., Relation between soy-associated isoflavones and LDL and HDL cholesterol concentrations in humans: a meta-analysis, *Eur. J. Clin. Nutr.*, 57, 940–946, 2003.
52. Law, M.R., Wald, N.J., and Thompson, S.G., By how much and how quickly does reduction in serum cholesterol concentration lower risk of ischaemic heart disease?, *Br. Med. J.*, 308, 367–372, 1994.
53. Law, M.R., Wald, N.J., Wu, T., Hackshaw, A., and Bailey, A., Systematic underestimation of association between serum cholesterol concentration and ischaemic heart disease in observational studies: data from the BUPA study, *Br. Med. J.*, 308, 363–366, 1994.
54. Jenkins, D.J., Kendall, C.W., Marchie, A. et al., Effects of a dietary portfolio of cholesterol-lowering foods vs. lovastatin on serum lipids and C-reactive protein, *JAMA*, 290, 502–510, 2003.
55. Messina, M.J., Potential public health implications of the hypocholesterolemic effects of soy protein, *Nutrition*, 19, 280–281, 2003.
56. Zhuo, X.G., Melby, M.K., and Watanabe, S., Soy isoflavone intake lowers serum LDL cholesterol: a meta-analysis of 8 randomized controlled trials in humans, *J. Nutr.*, 134, 2395–400, 2004.
57. Lovati, M.R., Manzoni, C., Gianazza, E. et al., Soy protein peptides regulate cholesterol homeostasis in Hep G2 cells, *J. Nutr.*, 130, 2543–2549, 2000.
58. Manzoni, C., Duranti, M., Eberini, I. et al., Subcellular localization of soybean 7S globulin in HepG2 cells and LDL receptor up-regulation by its alpha constituent subunit, *J. Nutr.*, 133, 2149–2155, 2003.
59. Squadrito, F., Altavilla, D., Squadrito, G. et al., Genistein supplementation and estrogen replacement therapy improve endothelial dysfunction induced by ovariectomy in rats, *Cardiovasc. Res.*, 45, 454–462, 2000.
60. Nestel, P.J., Yamashita, T., Sasahara, T. et al., Soy isoflavones improve systemic arterial compliance but not plasma lipids in menopausal and perimenopausal women, *Arterioscler. Thromb. Vasc. Biol.*, 17, 3392–3398, 1997.
61. Bonetti, P.O., Lerman, L.O., and Lerman, A., Endothelial dysfunction: a marker of atherosclerotic risk, *Arterioscler. Thromb. Vasc. Biol.*, 23, 168–175, 2003.
62. Herrington, D.M., Brown, W.V., Mosca, L. et al., Relationship between arterial stiffness and subclinical aortic atherosclerosis, *Circulation*, 110, 432–437, 2004.
63. Rimbach, G., De Pascual-Teresa, S., Ewins, B.A. et al., Antioxidant and free radical scavenging activity of isoflavone metabolites, *Xenobiotica*, 33, 913–925, 2003.
64. Wiseman, H., O'Reilly, J.D., Adlercreutz, H. et al., Isoflavone phytoestrogens consumed in soy decrease F(2)-isoprostane concentrations and increase resistance of low-density lipoprotein to oxidation in humans, *Am. J. Clin. Nutr.*, 72, 395–400, 2000.
65. Tikkanen, M.J., Wahala, K., Ojala, S., Vihma, V., and Adlercreutz, H., Effect of soybean phytoestrogen intake on low density lipoprotein oxidation resistance, *Proc. Natl. Acad. Sci. USA*, 95, 3106–3110, 1998.

66. Schoene, N.W. and Guidry, C.A., Genistein inhibits reactive oxygen species (ROS) production, shape change, and aggregation in rat platelets, *Nutr. Res.*, 20, 47–57, 2000.

67. Zhang, X., Shu, X.O., Gao, Y.T. et al., Soy food consumption is associated with lower risk of coronary heart disease in Chinese women, *J. Nutr.*, 133, 2874–2878, 2003.

68. Nagata, C., Takatsuka, N., and Shimizu, H., Soy and fish oil intake and mortality in a Japanese community, *Am. J. Epidemiol.*, 156, 824–831, 2002.

69. Sasazuki, S., Case-control study of nonfatal myocardial infarction in relation to selected foods in Japanese men and women, *Jpn. Circ. J.*, 65, 200–206, 2001.

70. Nagata, C., Ecological study of the association between soy product intake and mortality from cancer and heart disease in Japan, *Int. J. Epidemiol.*, 29, 832–836, 2000.

71. Kronenberg, F., Hot flashes: epidemiology and physiology, *Ann. N.Y. Acad. Sci.*, 592, 52–86, 123–133, 1990.

72. Nagata, C., Takatsuka, N., Kawakami, N., and Shimizu, H., Soy product intake and hot flashes in Japanese women: results from a community-based prospective study, *Am. J. Epidemiol.*, 153, 790–793, 2001.

73. Obermeyer, C.M., Menopause across cultures: a review of the evidence, *Menopause*, 7, 184–192, 2000.

74. Krebs, E.E., Ensrud, K.E., MacDonald, R., and Wilt, T.J., Phytoestrogens for treatment of menopausal symptoms: a systematic review, *Obstet. Gynecol.*, 104, 824–836, 2004.

75. Munro, I.C., Harwood, M., Hlywka, J.J. et al., Soy isoflavones: a safety review, *Nutr. Rev.*, 61, 1–33, 2003.

76. Ju, Y.H., Allred, C.D., Allred, K.F., Karko, K.L., Doerge, D.R., and Helferich, W.G., Physiological concentrations of dietary genistein dose-dependently stimulate growth of estrogen-dependent human breast cancer (mcf-7) tumors implanted in athymic nude mice, *J. Nutr.*, 131, 2957–2962, 2001.

77. Allred, C.D., Allred, K.F., Ju, Y.H., Virant, S.M., and Helferich, W.G., Soy diets containing varying amounts of genistein stimulate growth of estrogen-dependent (mcf-7) tumors in a dose-dependent manner, *Cancer Res.*, 61, 5045–5050, 2001.

78. Teede, H.J., Dalais, F.S., and McGrath, B.P., Dietary soy containing phytoestrogens does not have detectable estrogenic effects on hepatic protein synthesis in postmenopausal women, *Am. J. Clin. Nutr.*, 79, 396–401, 2004.

79. Atkinson, C., Warren, R.M., Sala, E. et al., Red-clover-derived isoflavones and mammographic breast density: a double-blind, randomized, placebo-controlled trial, *Breast Cancer Res.*, 6, R170–R179, 2004.

80. Chen, J., Hui, E., Ip, T., and Thompson, L.U., Dietary flaxseed enhances the inhibitory effect of tamoxifen on the growth of estrogen-dependent human breast cancer (mcf-7) in nude mice, *Clin. Cancer Res.*, 10, 7703–7711, 2004.

81. Bruce, B., Messina, M.J., and Spiller, G.A., Isoflavone supplements do not affect thyroid function in iodine-replete postmenopausal women, *J. Med. Food*, 6, 309–316, 2003.

82. Conrad, S.C., Chiu, H., and Silverman, B.L., Soy formula complicates management of congenital hypothyroidism, *Arch. Dis. Child*, 89, 37–40, 2004.

3

Processing of Soybeans into Ingredients

Mian N. Riaz

CONTENTS

Introduction

For at least 4000 years, soybeans and soybean products have been used extensively in the Orient as a source of protein and energy in human diets in one form or another. Soybeans are by far the most important crop in the world and are grown for a variety of agriculture and industrial uses. Over the past several decades, soybeans have become an increasingly important agricultural commodity and in the future should experience steady growth in production throughout the world. Consumers continue to demand healthier food products and are looking for positive health benefits, not just products that are less harmful. Given the choice, most consumers would prefer to enhance or improve their diet by means of food- and drink-related products rather than the more traditional pills, tablets, capsules, etc.[1] Recently, soy health benefits have been receiving greater coverage in the media. At the same time, more health professionals are recommending soy products to their patients.

Ten years ago, the average consumer was not aware of the health benefits of soy. Word gets around fast in the consumer food press community, though, and right now the word on nearly everyone's lips is "soy." Not long ago, soy had a negative connotation among manufacturers, perhaps due to the public's perception of soy as having an unpleasant taste and smell or the lack of scientific evidence and consumer awareness about the nutritional values of soy.

Traditional soyfoods, such as tofu, soymilk, miso, and tempeh, have been consumed for centuries by Asian populations. Soy proteins are acceptable in almost all diets, as they contain virtually no cholesterol and are lactose free, and they provide consumers with cholesterol-free, lower fat alternatives to animal protein. Both isolated and concentrated soy proteins are easily digested by humans and equal the protein quality of milk, meat, and eggs. Soy protein supplies all nine essential amino acids and offers many functional benefits to food processors. Modern soy products such as soy flour, concentrates, isolates, and textured soy protein have been used for several decades as functional ingredients by the food industries in the United States and Europe. The two categories of functional foods are (1) potential functional foods, and (2) established functional foods; soybeans are in the category of established functional foods.[2]

The increased acceptance of soy protein is due to its versatile qualities and good functional properties in food application. Several food industries are working to bring soyfoods to consumers rather than depending on consumers to seek them out. Conventional breads, snacks and crackers, and breakfast cereals to which soy has been added are likely to be particularly attractive (several breads containing soy have already proven to be successful). Over the last two decades, researchers and scientists all over the world have documented the health benefits of soy protein. Soy protein has been found helpful in reducing cholesterol levels, the effects of menopause, and the risks for several diseases (e.g., cancer, heart disease, osteoporosis).

Soybeans occupy a unique position among legumes. Traditional food legumes are rich in protein but contain a limited amount of oil (energy). Soybeans contain more protein than other legumes and a much higher content of edible oil, but raw soybeans cannot be used effectively in human foods unless processed properly. Several antinutritional factors have been identified in raw soybeans, including trypsin and chymotrypsin inhibitors, phytohemagglutinins (lectin), and allergenic factors such as glycinin and beta-conglycinin. These antinutritional factors can be deactivated, modified, or reduced through proper heat treatment to minimize or eliminate their adverse effects. Because all of these inhibitors are proteins, caution must be taken to ensure that no destruction of the oilseed protein occurs. This can be accomplished only through optimum processing and good quality control measures.[3]

From the nutraceutical and functional point of view, soy flour, soy concentrates, and soy isolates are the best way to add soy protein to the diet. In general, soy flour contains 50 to 54% protein, soy concentrates contain 65 to 70% protein, and soy isolates contain more than 90% protein. The nutraceutical and food industries use all three forms of soy products in their formulation. This chapter discusses the various processing methods used to produce the different kinds of soy flour, soy concentrates, soy isolates, soy germ, and soy fiber from whole soybeans; it does not address any oil-related soy products.

Processing of Soybeans

Whole soybeans can be processed into several different soy ingredients based on their application. Soybean ingredients that are currently available in the market for food applications or can be processed include: (1) roasted soy nuts, (2) enzyme-active full-fat soy flour and grits, (3) enzyme-inactive full-fat soy flour and grits (also called toasted full-fat soy flour and grits or heat-treated full-fat soy flour), (4) extruder-processed full-fat soy flour and grits (enzyme-inactive), (5) enzyme-inactive low-fat soy flour or grits, (6) enzyme-active defatted flake/flour with a 90 protein dispersibility index (PDI), (7) defatted soy flake/flour with a 70 PDI, (8) defatted soy flake/flour with a 20 PDI, (9) lecithinated soy flour, (10) textured soy flour, (11) refatted soy flour, (12) soy concentrates, (13) textured soy concentrates, (14) soy isolates, (15) soy germ, (16) chemically extracted soy isoflavones, (17) mechanically extracted soy isoflavones, (18) soy fiber from hulls, (19) soy fiber from cotyledon, and (20) organic soy flour and concentrates.[4]

Roasted Soy Nuts

Soybeans are cleaned, sized, color sorted, and washed prior to being soaked. The beans are evenly distributed into soaking tanks that have monitored water timers. The beans are soaked for a predetermined time based on the ambient temperature, moisture levels, etc. and then drained. The cleaned, soaked, and drained beans are dumped on a conveyor that carries the beans from the processing room to the roasting line. The beans are placed in a large hopper and then distributed on the stainless steel roasting belt by means of a vibratory feeder. The beans are spread evenly over the roasting belt with a second belt resting on top of them. This keeps the beans completely submerged in the soy oil used for roasting. The beans move through the roaster for a prescribed time based on, for example, their size, temperature, and moisture level.

After the beans have been roasted, they enter a second stainless steel conveyor that cools them. They move slowly through the cool-down conveyor while large fans pull the heat from them. The beans exit onto a shaker table that eliminates any loosened hulls or pieces. They are then conveyed into a flavoring vessel. Once the flavoring vessel has reached its prescribed volume, the required amount of flavoring is added. The vessel is then sealed and put into the roller/tumbler for mixing. After the flavoring step is completed, the vessel is opened and placed in the dumper. The dumper lifts the vessel and dumps the flavored, roasted soy nuts onto a conveyor that transports them to holding hoppers. The selected flavor are then directly packaged into either bulk or individual bags at the appropriate time. Soy nuts are ideal for healthy eating and dieting. These nuts are low in sodium and fat and rich in protein and fiber. These nuts can be mixed with other nuts or sold simply as soy nuts. A flow diagram for the production of roasted soy nuts is provided in Figure 3.1.

Enzyme-Active Full-Fat Soy Flour and Grits

Cleaning

Cleaning is critical for producing high-quality end products and protecting the processing equipment. Soybeans are cleaned on sieves under air aspiration to separate out dust, plant tissue, pebbles, and other lightweight contaminating material, as well as bigger impurities (e.g., stones, stems, nails). The larger impurities are primarily separated in destoners and magnetic iron separators. Cleaned beans are weighed with automatic hopper scales to

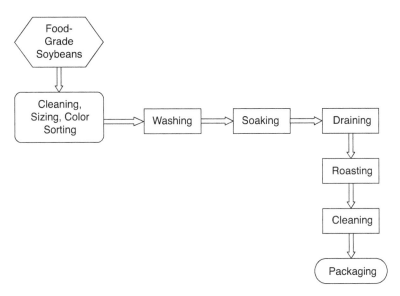

FIGURE 3.1
Production of roasted soy nuts.

provide a means to control the rate of feed and to record the total amount of raw material for accounting purposes. The next step is drying the soybeans.

Drying

Soybeans are dried in grain driers to a moisture level of 10%. The temperature of the dryer should be between 70 and 76°C to achieve the desired moisture level. Uniform drying of the soybeans is very important for removal of hulls, as it is necessary to remove the hulls from every individual bean, not just an average number of them. To do a good job of hull removal, it is essential to subject the beans to some kind of thermal impact. The hulls are attached to the bean with a proteinatious material, which, when exposed to heat, releases the hulls. After drying, the beans are stored for tempering for approximately 72 hours to stabilize the moisture content. Dried beans are usually cleaned to remove as many loose hulls, pods, sticks, and other foreign materials as possible. This step improves the efficiency of the cracking rolls and aspirator. The next step is cracking and dehulling the soybeans.

Cracking and Dehulling of Soybeans

The primary reason for dehulling is to produce high-protein end products. Hulls account for 7 to 8% of the volume of beans processed, and removing these hulls prior to processing increases the protein content of the desired end products. Clean and dry soybeans are cracked into suitable pieces for

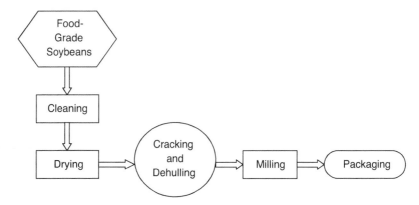

FIGURE 3.2
Production of enzyme-active full-fat soy flour and grits.

dehulling using a cracking mill. It is desirable to produce minimal fines and
no mashed beans. After cracking, the hulls are separated under air flow, and
the lighter hulls are withdrawn. It is important at this point to understand
exactly how air separation works, as it is currently the basis for all soybean
dehulling. With proper drying and cracking, the cotyledons should separate
easily from the hulls. As a practical matter, however, it is not possible to
obtain an absolutely complete separation of hulls from the cotyledons. The
next step is milling.

Milling

The dehulled soybeans are ground by hammer or impact pin mill into flours
of the desired particle sizes. Full-fat products are not easy to pulverize or
sieve. Coarse material can be separated from fine material by air classification. To make grits, a different size screen is used in the hammer mill. Grits
are used in bread for bleaching wheat flour and conditioning the dough.
Normal use in the United States is limited to 0.5% for white pan bread,
whereas in different parts of the world usage can be up to 2%. A flow diagram
for the production of enzyme-active full-fat soy flour and grits is provided
in Figure 3.2.

Enzyme-Inactive Full-Fat Soy Flour and Grits

This type of flour is also known as toasted full-fat soy flour or heat-treated
full-fat soy flour. In this process, soybeans are cleaned, as described earlier.
The soybeans are steamed under slight pressure for 20 to 30 minutes, then
cooled, dried, cracked to remove the hulls, and milled by means of sieving.
This type of flour has very low lipoxygenase activity and less beany flavor.
In this process, the undesirable enzymes have been destroyed by heating,
and the product has a low PDI (20 to 30). Various methods are available for

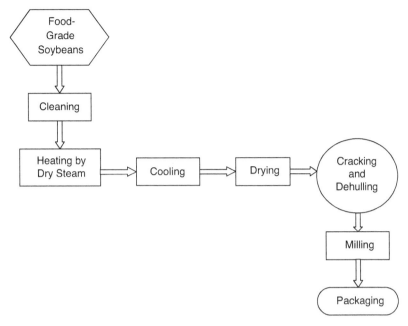

FIGURE 3.3
Production of enzyme-inactive full-fat soy flour and grits.

heating the soybeans, such as infrared radiation.[5] This type of full-fat soy flour is used in bakery products, soy breads, cakes and doughnuts, sweet goods, chocolates and confectioneries, spreads, and wafer fillings. A flow diagram for the production of enzyme-inactive full-fat soy flour and grits is provided in Figure 3.3.

Extruder-Processed Full-Fat Soy Flour and Grits (Enzyme-Inactive)

Extruder-processed full-fat soy flour is processed using wet or dry single-screw extruders. After cleaning, drying, and dehulling, the soybeans are coarse ground to be processed in the extruder. In this process, a single-screw extruder (wet or dry) is used to cook the soybeans. Extruded soybeans are then ground with an Alpine grinding mill to make a full-fat flour. This type of flour is used in soup or gravy and other foods in Brazil and Argentina. A flow diagram for the production of full-fat soy flour and grits by the extrusion process is provided in Figure 3.4.

Enzyme-Inactive Low-Fat Soy Flour or Grits

This type of flour also referred to as mechanically processed soybean flour. In this process, soybeans are cleaned, dried, dehulled, and cracked and then fed into a dry extruder. The dry extruder barrel is segmented but not jacketed;

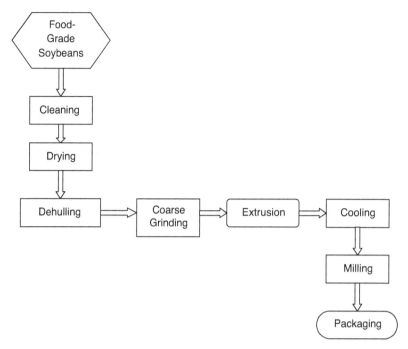

FIGURE 3.4
Production of extruder-processed full-fat soy flour and grits.

all of the heat of cooking is generated internally by friction. The soybeans
cook at 325°F (162°C) for 25 to 30 seconds. The extrudate exiting the extruder
is in a semifluid and frothy state, and the oil is free within the matrix. The
soybean extrudate is fed immediately into a screw press, where the oil is
pressed out. The protein-rich meal that comes out of the press in the form
of press cake is delumped using a roller mill and is passed through a meal
cooler.[6] After cooling, this material is ground into flour using a pulverizer.
This type of flour will have 6 to 8% oil, depending on the processing condition. This flour can be used in bread and other baked goods, as well as for
making textured soy protein. A flow diagram for the production of enzyme-
inactive low-fat soy flour and grits is provided in Figure 3.5.

Enzyme-Active Flake/Defatted Soy Flour or Grits (90 PDI)

This type of soy flake/flour is produced using a solvent extraction process.
After cleaning, drying, and dehulling, the next steps are conditioning and
flaking.

Conditioning

Cracked and dehulled soybeans are conditioned in steam cookers at approximately 70°C. Small amounts of water can be added to adjust the moisture

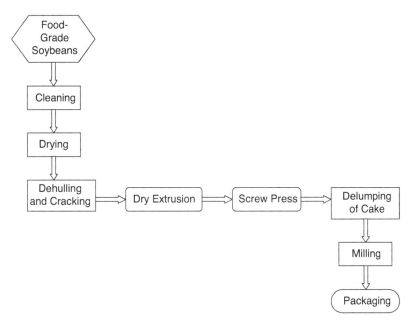

FIGURE 3.5
Mechanically expelled low-fat soy flour.

to approximately 11%, which makes the beans ideal for flaking. Correct heat treatment and adjustment of the moisture assist in disrupting the cell membranes during flaking so the oil can be extracted more easily. The main function of conditioning is to facilitate the flaking step.

Flaking

In order to extract the oil from the beans it is important to modify the cell structure of the beans. To achieve this, the beans are flaked in flaking machines, as solvent can flow much more readily through a bed of flakes than through a bed of soy meats or fine particles. The ideal thickness of a flake is from 0.25 to 0.35 mm. The thickness of these flakes depends on the size of the cracked beans, conditioning, and adjustment of the flaking rolls.

Extraction of Flakes

Soy flakes are fed into a flash desolventizer or vapor desolventizer. In the flash desolventizer, solvent wet flakes move directly from the extractor into superheated vapor blown at a high velocity through a long tube. Solvent is heated under pressure to 115 to 138°C. As this vapor is blown through the tube, flakes are conveyed in and picked up by the vapor stream. In a few seconds, the superheated solvent flashes off the residual liquid solvent, leaving desolventized flakes to be collected in the cyclone.[7] High PDI soy

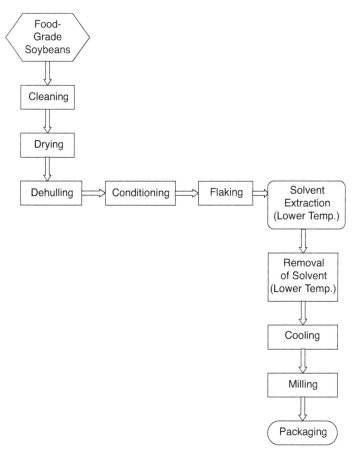

FIGURE 3.6
Production of enzyme-active defatted flake/flour with a 90 PDI.

ingredients are more soluble but have highly active enzymes and antinutritional factors. These flakes have a 90 PDI. Defatted flakes can be milled
using a hammer mill to produce soy flour. At least 97% of the flour must
pass through a U.S. Standard No. 100 sieve. This flour contains 52 to 54%
protein. This type of flour is used in bakery applications as a crumb whitener
or dough conditioner. The use of this flour is very limited (0.5%) because
of the enzymes and beany flavor. A flow diagram for the production of
enzyme-active flake/defatted soy flour or grits is provided in Figure 3.6.

Enzyme-Inactive Defatted Soy Flakes/Flour or Grits (70 PDI)

This type of soy flake/flour is produced using the same method as described
previously (enzyme-active), except that solvent extraction is done at a higher
temperature. This temperature inactivates the lipoxygenase enzymes and
produces defatted soy flake/flour that is enzyme inactive and has a PDI of

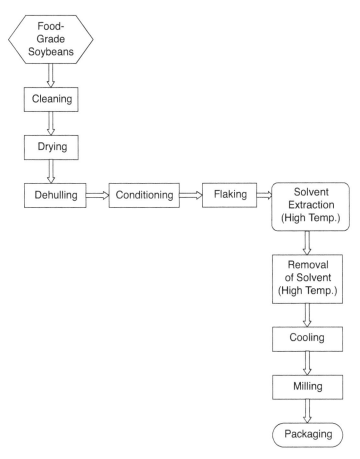

FIGURE 3.7
Production of defatted soy flake/flour with a 70 PDI.

70. Typical applications of this flake/flour include waffle and pancake mixes, bread, doughnuts, tortillas, and bagels. Its use in bakery products is limited to up to 5% of the dry ingredients. A flow diagram for the production of enzyme-inactive defatted soy flakes/flour or grits (70 PDI) is provided in Figure 3.7.

Enzyme-Inactive Defatted Soy Flakes/Flour or Grits (20 PDI)

This type of soy flour is produced using the same method as described above for the 70 PDI product, except this type of flake/flour is further toasted after the solvent extraction to reduce the PDI to 20. Typical applications for this type of flake/flour include various bakery applications (e.g., biscuits) and as a milk replacer; its use in bakery products is limited to up to 5% of the dry ingredients. A flow diagram for the production of enzyme-inactive defatted soy flakes/flour or grits (20 PDI) is provided in Figure 3.8.

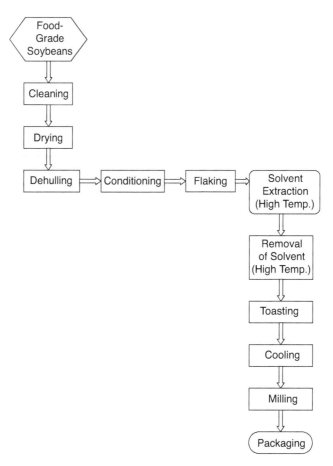

FIGURE 3.8
Production of defatted soy flake/flour with a 20 PDI.

Textured Soy Flour

Defatted soy flour of 70 PDI is processed through a thermoplastic extrusion process to produce textured soy flour. This type of flour is used in processed meats, pizza topping, hamburger patties, tacos, etc. The size of the textured flour will depend on the application and desired texture. After texturizing, the flour can be processed through a Comitrol mill to produce different sizes. A flow diagram for making textured soy flour is provided in Figure 3.9.

Lecithinated Soy Flour

This type of flour is produced by spraying lecithin and vegetable oil on 70 and 20 PDI soy flour. The most common rates are 3, 6, and 15% lecithin by weight. This type of flour is a good emulsifier and can be used for partial

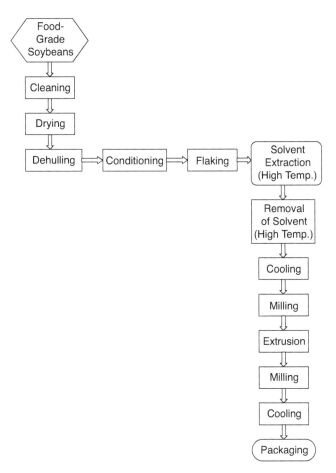

FIGURE 3.9
Production of lecithinated soy flour.

or total replacement of egg. It improves the dispersion of the flour and other admixed ingredients in confections and cold beverage products.[5] It can also improve water retention in baked good and extend their shelf lives. Also, this type of flour is a good animal protein substitute in vegetarian cookies. A flow diagram for the production of lecithinated soy flour is provided in Figure 3.10.

Refatted Soy Flour

This flour is produced by adding 1 to 15% fat back into the flour for certain applications, primarily to reduce the dustiness and provide fat for a product formula.[5] A flow diagram for the production of refatted soy flour is provided in Figure 3.11.

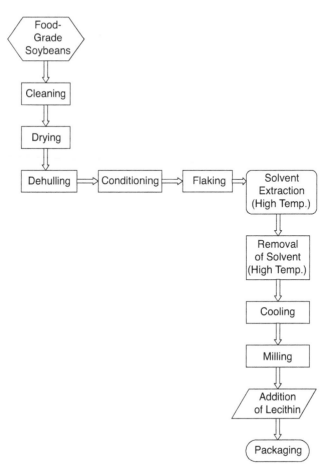

FIGURE 3.10
Production of textured soy flour.

Production of Soy Protein Concentrates

Soy protein concentrates contain at least 65% protein. They are processed selectively by removing the soluble carbohydrates from soy protein flour using aqueous alcohol or isoelectric leaching. Defatted flakes or soy flour are used as the starting material. When processing soy concentrates, the objective is to immobilize the protein while leaching away the soluble carbohydrates and removing the strong flavor components and flatulence sugars (stachyose and raffinose). In turn, the percentage of protein and dietary fiber is increased. Several different methods are used to produce soy concentrates, including:[5]

- Extraction of flakes with aqueous 20 to 80% ethyl alcohol
- Acid leaching of flakes or flour
- Denaturing the protein with moist heat and extraction with water[8]

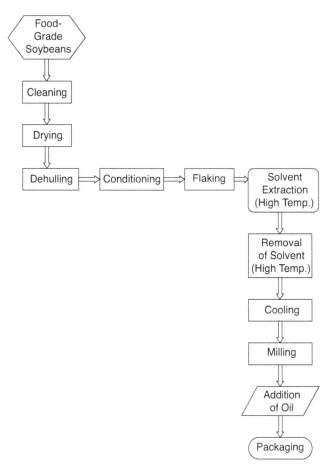

FIGURE 3.11
Production of refatted soy flour.

Alcohol extraction is considered to produce the blandest products. Mild heat drying conditions are used in an acidic water extraction process to maintain a high PDI.[9] The isoflavone content in soy concentrates depends on whether it has been water washed or alcohol washed, as isoflavones are soluble in alcohol. Because of their more bland flavor, soy concentrates are often preferred over soy flour. Typically, soy concentrates are used in meat patties, pizza toppings, bakery products, and meat sauces. A flow diagram for the production of soy protein concentrates is provided in Figure 3.12.

Textured Soy Protein Concentrates

Soy protein concentrates are texturized by using a wet extruder. The size of the textured concentrates depends on the application. After extrusion, textured soy protein concentrates can be converted into various sizes using

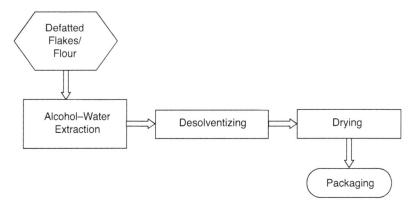

FIGURE 3.12
Production of soy concentrates.

size reduction equipment such as the Comitrol mill. Typical applications of textured soy protein concentrates include their use in meat products, such as patties, sausages, and pizza topping. A flow diagram for the production of textured soy protein concentrates is provided in Figure 3.13.

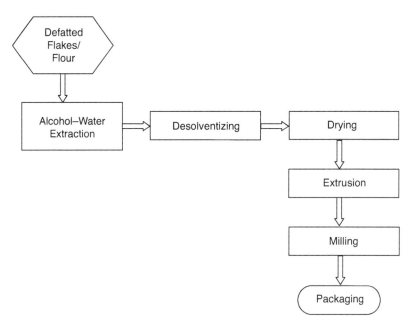

FIGURE 3.13
Production of textured soy concentrates.

Production of Soy Protein Isolates

Several methods for manufacturing isolated soy protein have been developed, but the only commercial procedure currently being used is extraction of defatted soy flakes or soy flour with water, followed by centrifugation. In this process, the protein is solubilized at a pH of 6.8 to 10 at 27 to 66°C by using sodium hydroxide and other alkaline agents approved for food uses. The protein solution is then separated from the flakes or flour by centrifugation. The solids are recovered as byproducts containing 16 to 36% protein, 9 to 13% crude fiber, and 45 to 75% total dietary fiber when dried to a moisture content of 6 to 7%. The solution is then acidified to pH 4.5 by using hydrochloric or phosphoric acid, and the protein is precipitated as a curd. The curd is then washed with water and concentrated by centrifugation; it can then be neutralized to pH 6.5 to 7.0 or spray dried in its acidic form.[10] Soy isolates, the most concentrated form of soy protein (90% dry bases) available, generally contain about half as many isoflavones as soy flour as a result of losses during processing. Functionally, isolated soy proteins have some of the same attributes as soy flour and concentrates but are normally utilized to take advantage of their higher solubility and whipability, as well as their foaming and binding functionality for a number of food products, including meat. A flow diagram for the production of soy protein isolates is provided in Figure 3.14.

Soy Germ

Soy germ is mechanically separated from soybeans during the process of dehulling and cracking. Several companies use patented technology to separate the germ, which is then roasted, milled, and converted into various forms, such as flour and granules of different sizes. The germ is considered to be a very rich source of isoflavones and other nutrients. On average, soy germ will contain 40% protein, essential fatty acids, omega 3 and 6, lecithin, vitamin E, and saponin.[11] Soy germ is by far the most nutrient-dense soy ingredient one can consume to obtain all of the proven health benefits of the soybean. A flow diagram for the production of soy germ is provided in Figure 3.15.

Isoflavones

Isoflavones are found in high amounts in soybeans. Genistein and daidzein are the major isoflavones found in soybeans. In whole soybeans, the isoflavones are present at a concentration of 0.2 to 3.0 mg/g. Soy experts recommend consuming 30 to 50 mg of isoflavones per day. To do so, one would have to consume 20 to 150 g of soyfoods, 200 g or almost 7 oz. of tofu, or 3 cups of soymilk daily. Isoflavones are available in several different forms: crunchy granules, flour, or fine powder. Soy isoflavone concentrates can be incorporated in cereals, snacks, cookies, breads, health bars, drinks, and meat replacements.

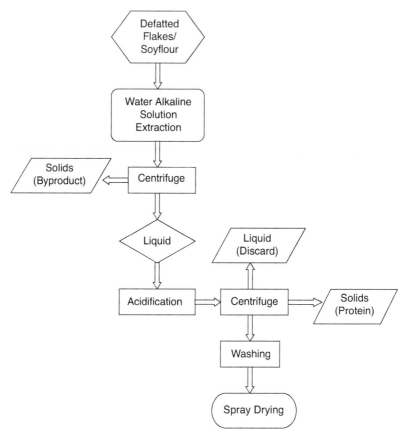

FIGURE 3.14
Production of soy isolates.

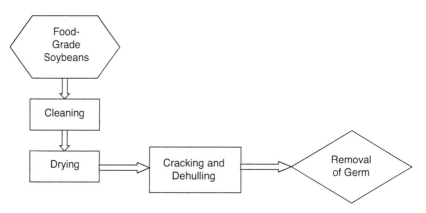

FIGURE 3.15
Production of soy germ.

Isoflavone Processing

The concentration of isoflavones in soyfoods and soy ingredients depends on: (1) the genetics of the soybeans, (2) the environment in which the soybeans were grown, and (3) the processing technologies used. The level of isoflavones in soybeans grown in the same field in the same year may vary in concentration by a factor of five. Two methods are used to extract isoflavones from soybeans: chemical extraction and mechanical extraction.

Chemical Extraction of Isoflavones

Material from the solvent extraction process, when converted into soy concentrates and isolates, generates a byproduct referred to as soy molasses. This molasses is a rich source of isoflavones. In the conventional process for the production of soy protein concentrates in which soy flakes or flour are extracted with an aqueous acid or an aqueous alcohol to remove water-soluble materials from the soy flakes or flour, a large portion of the isoflavones is solubilized in the extract. The extract of water-soluble material that includes the isoflavones is the soy molasses. The isoflavones are chemically extracted by a number of different patented processes. A flow diagram for the production of isoflavones using chemical extraction is provided in Figure 3.16.

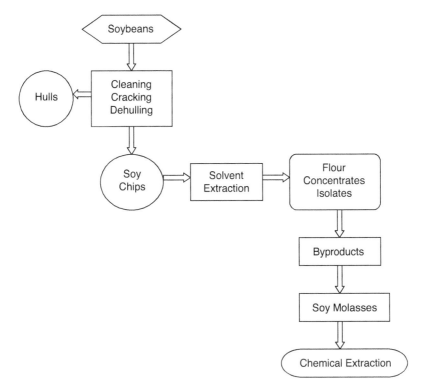

FIGURE 3.16
Production of chemically extracted soy isoflavones.

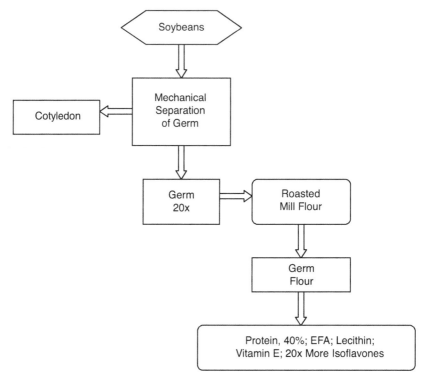

FIGURE 3.17
Production of mechanically extracted soy isoflavones.

Mechanical Extraction of Isoflavones

In the mechanical extraction of isoflavones, soybean germ is mechanically separated from the cotyledons, as discussed earlier. The germ contains 5 to 6 times more isoflavones than the rest of the bean. On average, 400 pounds of soybeans will produce 1 pound of germ. Some companies sell soy germ as a source of mechanically extracted isoflavones to be used in food applications, such as cereal, bars, supplements, and beverages. A flow diagram for the production of isoflavones by means of mechanical processing is provided in Figure 3.17.

Soy Fiber

Two types of edible fiber are produced from soybean processing operations: hulls and soy cotyledon fiber.

Soy Hull as a Source of Fiber

Soy fiber (also known as soy bran) is produced from the hull (or seed coat) of the soybean. On a dry weight basis, hulls constitute about 8% of the total seed, depending on variety and the seed size. In general, the larger the seed, the lower the proportion of hulls. The hull of a dry mature soybean contains about 7 to 8% moisture. Dry soy hulls contain approximately 85.7% carbohydrates, 9% protein, 4.3% ash, and 1% lipids.[12] Kikuchi et al.[13] studied the chemical and physical properties of soybean cell wall polysaccharides and the changes that occur during cooking. They found that the soy cell wall contains about 30% pectin, 50% hemicellulose, and 20% cellulose; therefore, most soybean carbohydrates fall into the general category of dietary fiber. The outer hull is an excellent source of dietary fiber (6 g fiber per 1 cup cooked). Soy fiber contains approximately 38% crude fiber and can have a total dietary fiber content as high as 76%.[14] Fiber components in soybean hulls and in the cotyledon are distinct and should be considered separately.

After separation from the seeds during rolling or flaking, the hulls are toasted and ground, then blended back into defatted soy meal to make a meal containing 44% protein. Recently, several new uses of soy hulls have been explored, with an emphasis on human food applications. During soybean processing operations, the hulls are removed, washed, sterilized, dried, toasted, and ground to different sizes depending on the desired application, such as for use in breads, cereals, and snacks. The natural grittiness of the product typically requires fine grinding.

Following are specifications for soy hull fiber available in the U.S. market:[5]

- Total dietary fiber — 92.0%
- Moisture — 3.5%
- Fat — 0.5%
- Protein — 1.5%
- Ash — 2.5%
- Caloric content — 0.1 Kcal/g
- Water absorption — 350 to 400%
- pH — 6.57 to 7.5

Sieve analyses are as follows:

- U.S. no. 80 — 0%
- U.S. no.100 — 0%
- U.S. no. 140 — 2%
- U.S. no. 200 — 7%
- Through U.S. no. 200 — 91%

A flow diagram for the production of soy fiber from hulls is provided in Figure 3.18.

Soy Cotyledon Fiber

During processing, beans are carefully selected for color and size. These beans are then cleaned and conditioned (to make removal of hulls easier), cracked and dehulled, and rolled into flakes. The flakes are subjected to a

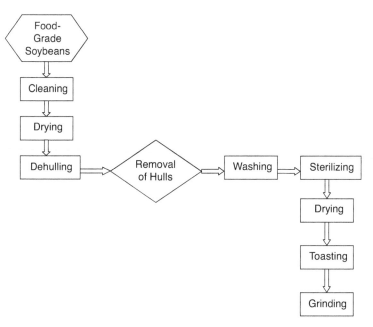

FIGURE 3.18
Production of fiber from hulls.

solvent bath that removes the oil. The solvent is removed and the flakes are dried, creating the defatted soy flakes that are the basis of all soy protein products. The production of soy protein isolate involves the removal of soluble material from defatted flakes by aqueous leaching. In this process, part or all of the protein fraction is solubilized, usually at an alkaline pH, and the insoluble protein is separated by centrifugation. These solids are analogous to the okara product resulting from the production of soymilk or tofu, but they differ in that they do not contain hulls and have been defatted and treated with mild alkali. The insoluble residue obtained during protein isolation is made up mainly of various hemicellulose materials. After appropriate purification, this material is dried and sold as soy fiber, in competition with other fibers, such as wheat or oat bran. This soy fiber contains 80% total dietary fiber. The dietary fiber content of soy fiber varies greatly, depending on the method used for analysis. Following are manu-facturers' specifications for soy hulls sold in the United States:[5]

- Dietary fiber — 75% on a moisture-free basis (65% noncellulosic and 10% cellulosic polysaccharides)
- Moisture — 12%
- Fat — 0.2%
- Ash — 4.5%

Several of these products are made during the processing of soy isolates, after dehulling, oil extraction, and removal of the soluble sugars and proteins. Using proprietary processing techniques, the water adsorption or binding can be controlled to make products useful in, for example, beverages, puddings, or soups or in food systems with less water, such as baked breads, muffins, crackers, cookies, or breakfast bars. A variety of soy fibers can be produced and are available for commercial application. Most of the published research is on polysaccharides, the cellulosic and noncellulosic structural components of the inner cell wall.[15] Generally, it is thought that soluble fiber is associated with cholesterol lowering and improved diabetic control, whereas insoluble fiber is associated with enhanced bowel function. Although soy polysaccharides contain primarily insoluble dietary fiber, clinical studies have shown it to have properties of both insoluble and soluble fiber. Soy protein concentrates contain both insoluble and soluble fiber; therefore, foods made with soy protein concentrates can be good sources of dietary fiber. Unlike meat, milk, fish, and eggs, a bounty of protein is combined with dietary fiber in soy products, such as soy flour and textured vegetable protein. A flow diagram for the production of fiber from cotyledon is provided in Figure 3.19.

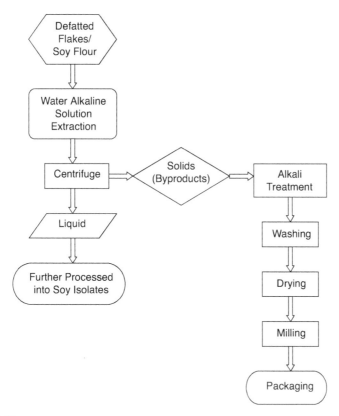

FIGURE 3.19
Production of soy fiber from cotyledon.

Organic Soy Flour and Concentrates

Organic soy flour and concentrates that have recently become available for food applications are made using solvent-free proprietary and patented methods, but not enough information is available to discuss these processes here.

References

1. Riaz, M.N., Processing of soybeans into different functional ingredients for food application, in *Proceedings of the ASA Soyfood Symposium Series*, Singapore, American Soybean Association, St. Louis, MO, 2001, pp. 5–12.
2. Riaz, M.N., From soy processing to functional foods and nutraceutical, in *Soy and Health 2000: Clinical Evidence, Dietetic Application*, Descheemaeker, K. and Debruyne, I., Eds., Garant, Belgium, 2001, pp. 37–45.
3. Riaz, M.N., Production of fullfat soy by extrusion, in *Proceedings of the Third International Soybean Processing and Utilization Conference (ISPUC-III)*, Korin Publishing, Japan, 2000, pp. 375–376.
4. Riaz, M.N., Soy processing, in *Soy Ingredients and Application Manual*, AACC/IFT, Minneapolis, MN, 2002.
5. Lusas, E.W. and Rhee, K.C., Soy protein processing and utilization, in *Practical Handbook of Soybean Processing and Utilization*, Erickson, D. R., Ed., AOCS Press, Champaign, IL, 1995, pp. 117–160.
6. Wijeratene, W., Making low-fat soy flour for texturized vegetable protein, in *Manual of Texturized Vegetable Protein and Other Soy Products*, Riaz, M.N. and Barron, M., Eds., Texas A&M University, College Station, 2004.
7. Fulmer, R.W., The preparation and properties of defatted soy flour and their products, in *Proceedings of the World Congress on Vegetable Protein Utilization in Human Foods and Animal Feedstuffs*, AOCS Press, Champaign, IL, 1988, pp. 55–60.
8. Ohren, J.A., Process and product characteristics for soya concentrates and isolates, *J. AOCS*, 58, 333, 1981.
9. Lusas, E.W. and Riaz, M.N., Soy protein products: processing and use, *J. Nutr.*, 125, 573S–580S, 1995.
10. Johnson, D.W. and Kikuchi, S., Processing for producing soy protein isolates, in *Proceedings of the World Congress on Vegetable Protein Utilization in Human Foods and Animal Feedstuffs*, Applewhite, T.H., Ed., AOCS Press, Champaign, IL, 1989, pp. 226–240.
11. Riaz, M.N., Functional soyfoods, in *Manual of Texturized Vegetable Protein and Other Soy Products*, Riaz, M.N. and Barron, M., Eds., Texas A&M University, College Station, 2004.
12. Liu, K., *Soybeans: Chemistry, Technology, and Utilization*, Aspen Publications, Gaithersburg, MD, 1999.
13. Kikuchi, T.S., Ishil, S., Fukushima, D., and Yokotsuka, T., Food chemical studies on soybean polysaccharides. I. Chemical and physical properties of soybean cell wall polysaccharides and their changes during cooking, *J. Agric. Chem. Soc.*, 45, 228, 1971.
14. *Soy Protein Products: Characteristic, Nutritional Aspect and Utilization*, Soy Protein Council, Washington, D.C., 1987.
15. Aspinall, G.O., Bigvie, R., and McKay, J.L., Polysaccharide component of soybeans, *Cereal Sci. Today*, 12, 224, 1967.

4

Soy Ingredients in Baking

M. Hikmet Boyacioglu

CONTENTS

Introduction

Bread is the staff of life, and it has been the cause of wars, peace, and revolutions throughout history. Dr. Norman Borlaug, Nobel Peace Price recipient in 1970, said, "If you desire peace, cultivate justice, but at the same time, cultivate the fields to produce more bread; otherwise, there will be no peace." Bread is eaten with every meal by every ethnic and age group every day of the year, regardless of economic level.[1] Over the last few decades, research in the area of baking technology has evolved according to the nutritional concerns, applications of new or improved techniques, and consumer trends at the time.[2] These trends have led us to use soy ingredients extensively in baking. Soybeans have been used as human food for nearly 5000 years, and the first soy protein food ingredients were soy flours, used primarily in breads and other bakery products in the United States in the 1940s. Since then, because of their excellent functional and nutritional characteristics, soy ingredients have been used in baking at an increasing rate due to the U.S. Food and Drug Administration's approval in 1999 of the health claim that the consumption of soy protein can reduce the risk of coronary heart diseases. This chapter discusses the utilization of soy ingredients in baking, emphasizing the types of soy ingredients, the functions of these soy ingredients, and the application of soy ingredients in bakery products. Because the term *baking* does not apply only to the production of bread, the chapter covers not only bread and variety breads but also cakes, pastries, cookies, biscuits, donuts, and special applications.

Types of Soy Ingredients in Baking

Soy ingredients that have unique functional and nutritional properties have found wide application in bakery products. These soy ingredients include soy flours, soy grits, soy protein concentrates, soy protein isolates, textured soy protein, soy brans, and soy germs. Soy flour can be defined as flour produced from soybean seeds that have been hulled and are then ground into a fine granulation. Soy flours can also be classified as defatted soy flour, enzyme-active soy flour, low-fat soy flour, high-fat soy flour, full-fat soy flour, and lecithinated soy flour. Defatted soy flour is a finely granulated product obtained by grinding hulled soybeans after oil extraction. Enzyme-active soy flour is a flour produced from defatted soybeans that have been processed in such a manner as to preserve the activity of the enzyme lipoxidase (lipoxygenase). Full-fat soy flour is obtained from hulled soybeans and retains all of the original fat content of the soybean. Lecithinated soy flour is soybean flour for which the normal lecithin content of 0.5 to 1.5% has been augmented to 15% in order to increase its emulsifying properties.[3] Soy grits are identical to soy flours, the only difference being a larger particle size; they are used to enhance the nutritional and textural quality of cookies, crackers, and specialty breads.[4] Soy protein concentrate is a protein produced by the extraction of sugars, soluble carbohydrate material, mineral matter, and other minor constituents from defatted soy flour.[3] Obtaining higher protein fortification levels is generally the only reason to use soy protein

TABLE 4.1

Proximate Composition of Some Soy Ingredients per 100-g Edible Portion

Ingredient	Moisture (g)	Protein (N × 6.25) (g)	Total Lipids (Fat) (g)	Total Dietary Fiber (g)	Ash (g)
Soy flour:					
Defatted	7.25	51.46	1.22	15.5	6.15
Low-fat	2.70	50.93	6.70	10.2	6.09
Full-fat	5.16	37.80	20.65	9.6	4.46
Soy protein concentrate (by acid wash)	5.80	63.63	0.46	5.5	4.70
Soy protein isolate (potassium type)	4.98	88.32	0.53	2.0	3.58

Source: USDA National Nutrient Database for Standard Reference, Release 17, Nutrient Data Laboratory (http://www.nal.usda.gov/fnic/foodcomp/search/).

concentrates in bakery products; nutritionally and functionally, soy flours do the same job more economically.[5] Soy protein isolates are proteins separated from soybeans by various methods and often modified chemically or enzymatically to impart certain desirable functional properties. Soy bran is produced by toasting and grinding the seed coat portion of the soybean.[3,6] Soy germ, which comprises only 2% of the total soybean, can be used in baked and extruded products and is widely used in cereal-based products as an adjunct to other soy ingredients to increase overall isoflavone levels. Because of the relatively high proportion of isoflavones found within soy germ, inclusion rates are typically 1 to 2% of the total formulation of the product.[7]

Recently, the first soy flour to be recognized as whole grain, according to the definition of the American Association of Cereal Chemists (AACC) for whole grains, was introduced in the United States. The new whole-grain soy flour features 80% more dietary fiber than whole wheat and can be used at inclusion levels that support claims of the product being a "good source" or "excellent source" of FDA protein and fiber.[8] Table 4.1 provides the composition of soy ingredients used in bakery products. Soy protein products (SPPs) are also covered by a Codex General Standard.[9] According to the standard, soy protein products must conform to the essential compositional requirements shown in Table 4.2.

Functions of Soy Ingredients in Baking

The functional properties of proteins are a result of their physicochemical properties and their interactions with other food components. Functional properties are important not only in determining the quality of the final

TABLE 4.2

Codex General Standard for Soy Protein Products

Component	Soy Protein Flour (% dry weight basis)	Soy Protein Concentrate (%)	Soy Protein Isolate (%)
Moisture (maximum)	10	10	10
Crude protein ($N \times 6.25$)	50 to <60	65 to <90	90
Ash (maximum)	8	8	8
Crude fiber (maximum)	5	6	0.5

Source: CAC, *Codex General Standard for Soy Protein Products*, Codex Standard 175-1989, Codex Alimentarius Commission, Rome.

product but also in facilitating processing (e.g., improved machinability of cookie dough). In order to utilize soy ingredients effectively, food processors should have detailed information regarding the methods of preparation and processing of the soy products they are using because these affect the composition and functional properties of the component proteins.[10] Although soy ingredients have many functional properties in food systems, only the effects of soy protein preparations in bakery products are addressed in this chapter (Table 4.3).

Solubility

Solubility is one of the most basic physical properties of proteins and a prime requirement for any functional application. To obtain optimum functionality in uses where gelation, solubility, emulsifying activity, foaming, and lipoxygenase activity are required, a highly soluble protein is required. Soluble protein preparations are also easier to incorporate into foods. Proteins with low solubility indices have limited functional properties and more limited uses. The solubility of soy protein is also significantly affected by the method used for its production. The extent of heat treatment during processing determines the use of soy proteins in bakery products (Table 4.4). Heat treatment, especially moist heat, rapidly insolubilizes soy proteins; however, heat treatment is necessary to desolventize, to inactivate antinutrient compounds, and to improve the flavor of soy flours. Nonheated soy flours, while possessing high lipoxygenase activity, have a bitter, beany flavor and limited applications. To balance enzyme activity, flavor quality, and solubility criteria, processors produce defatted soy flours with a range of solubilities. Concentrates and isolates are prepared from minimally heat-treated flours and generally possess good solubility.

Protein solubility is a measure of the percentage of total protein that is soluble in water under controlled conditions and is a measure of the degree of heat treatment to which the soy flake has been subjected. The protein solubility is closely related to the functional properties for bakery food

TABLE 4.3

Selected Functional Characteristics of Soy Protein in Baking Systems

Functional Property	Mode of Action	Baking System	Protein Form Used
Emulsification			
Formation	Formation and stabilization of fat emulsions	Breads, cakes	Flour, concentrates, isolates
Fat Adsorption			
Prevention	Binding of free fat	Donuts, pancakes	Flour, concentrates
Water Absorption and Binding			
Uptake	Hydrogen bonding of water, entrapment of water, no drip	Breads, cakes	Flour, concentrates
Retention	Hydrogen bonding of water, entrapment of water, no drip	Breads, cakes	Flour, concentrates
Dough formation	—	Baked goods	Flour, concentrates, Isolates
Cohesion–adhesion	Protein acts as adhesive material	Baked goods	Flour, concentrates, isolates
Elasticity	Disulfide links in deformable gels	Baked goods	Flour, isolates
Flavor-binding	Adsorption, entrapment, release	Baked goods	Concentrates, isolates, hydrolyzes
Foaming	Forms stable films to trap gas	Whipped toppings, chiffon desserts, angel cakes	Isolates, soy whey, hydrolyzes
Color Control			
Bleaching	Bleaching of lipoxygenase	Breads	Flour
Browning	Maillard, caramelization	Breads, pancakes, waffles	Flour

Source: Adapted from Kinselle, J.E., *J. AOCS*, 56, 242, 1979; Stauffer, C., *Milling & Baking News*, January 23, 2001.

TABLE 4.4

Soy Protein Solubility Requirement for Selected Bakery
Applications

Application	Protein Dispersibility Index (PDI)
Emulsifying, foaming	>90
Lipoxygenase bleaching of flour and bread	>85
Water absorption in bakery products	60
Waffles	30
Crackers, cereals	15

Source: Adapted from Kinselle, J.E., *J. AOCS*, 56, 242, 1979.

applications. Several methods are used to determine protein solubility:
nitrogen solubility index (NSI), the protein dispersibility index (PDI), and
the protein solubility index (PSI). Each of these tests indicates the percentage
of total soluble nitrogen in water, with a range of values from 0 to 100.[6] The
PDI decreases with higher levels of heat treatment. The more dispersible
types of soy flours (high NSI or PDI) are used in bakery and cereal products
and are added directly to the dough.[4] Enzyme-active soy flour has a minimum water solubility of 70%.[11] Soy flours subjected to minimal heat treatment (PDI 80) show high lipoxygenase activity and are used at 0.5% to
bleach flour and improve the flavor of bread. Flours with a PDI of approximately 60 are most commonly used (1 to 2% in bread, 10% in waffles and
pancakes), as they markedly improve the water-binding capacity of these
products (Table 4.4).[10,12]

Emulsification

The ability of protein to aid in the formation and stabilization of emulsions
is critical for many food applications, including cake batters. In general, the
emulsifying capacity of soy protein products increases with increasing solubility. Accordingly, soy proteins progressively reduce interfacial tension as
concentration is increased.[10,12]

Foaming

Foaming, the capacity of proteins to form stable foams with gas by forming
impervious protein films, is an important property in some food applications, including angel and sponge cakes. Soy protein exhibits foaming properties closely correlated to its solubility. Studies have revealed differences in
the foaming properties of various soy protein products, and soy isolates are
superior to soy flours and concentrates.[10,12]

Gelation

Protein gels are three-dimensional networks in which water is entrapped. Gels are characterized by a relatively high viscosity, plasticity, and elasticity. Whereas soy flour and concentrates form soft, fragile gels, soy isolates form firm, hard, resilient gels. Protein gelation is concentration dependent; a minimum of 8% protein concentration is necessary for soy isolates to form a gel. The general procedure for producing soy protein gel involves heating the protein solution at 80 to 90°C for 30 minutes followed by cooling at 4°C. The ability of the gel structure to provide a matrix to hold water, fat, flavor, sugar, and other food additives is very useful in a variety of food products.[10,12]

Water-Binding Capacity

Bound water includes all hydration water and some water loosely associated with protein molecules following centrifugation. The amount of bound water generally ranges from 30 to 50 g/100 g protein. Soy isolate has the highest water-binding capacity (about 35 g/100 g) because it has the highest protein content among soy protein products.[12] Soy concentrates contain polysaccharides, which absorb a significant amount of water. Processing conditions can affect the amount of water that can be absorbed. In fact, these conditions can be varied to influence how tightly the water is bound by the protein in the finished food product.[4] Soy proteins differ considerably from wheat protein in their chemical composition as well as in their physical properties due to their total lack of elasticity. The addition of soy proteins to wheat flour thus dilutes the gluten proteins and starch. Soy proteins exhibit a strong binding power that provides some resistance to dough expansion, the effect being somewhat proportional to the level of soy flour employed. This can be partially overcome by increasing the amount of water used in dough making and by a longer proofing time. The binding power of soy flour is closely related to its high water absorption capacity, which, in the case of the defatted product, is equivalent to 110% by weight. Hence, soy flour will absorb an amount of water equal to its weight when mixed with wheat flour to normal dough consistency. With full-fat flour, however, no measurable increase in dough absorption results from normal use levels of the soy product.[13]

Water-Holding Capacity

Water-holding capacity is a measure of entrapped water that includes both bound and hydrodynamic water. The literature reveals some variability in data regarding water-holding capacity which may have resulted from variations in the methods of determination. The water-holding capacities of soy flour, concentrate, and isolate were reported as 2.6, 2.75, and 6.25 g/g of solids, respectively. Because heat treatment enhances this functional property, soy flours with PDIs of 15, 55, 70, and 85 had water-holding capacities of 209, 307, 308, and 207 g water per g flour, respectively. The water-holding capacity of protein is very important because it affects the texture, juiciness, and taste. Also, the ability of soy protein to bind and retain water enhances the shelf life of bakery products.[10,12] All soy protein concentrates, regardless of the process used, do have certain fat- and water-holding characteristics.[4]

Color Control

Soy flour is rich in the enzyme lipoxygenase, which plays a major role in its bleaching action. Soy flour contains a type-1 (LOX-1) and a type-2 lipoxygenase (LOX-2), and earlier studies have shown that purified LOX-2, but not LOX-1, bleaches flour pigments. Lipoxygenase oxidizes carotenoid and chlorophyll pigments in flour to their colorless form, a bleaching action that whitens flour and bread crumbs; however, because LOX-2 lipoxygenase also oxidizes fatty acids, achieving a whiter flour increases the risk of rancidity. For this reason, the amount of enzyme-active flour is restricted to approximately 1% flour weight.[14–18]

Health Benefits of Soy Ingredients in Baking

The health benefits of soybeans have been recognized for millennia. Soyfoods and their isoflavones appear to have clear protective effects related to coronary heart disease and probable protective and therapeutic effects related to osteoporosis. The effects on the kidneys are clear, and these protective effects are under study. While the greatest interest may lie in their chemopreventive effects related to cancer, much more research is required. The effects of soyfoods on cognitive function are unclear and also require further research. The use for menopausal symptoms appears promising, and postmenopausal women who cannot or choose not to take hormone replacement therapy may be ideal candidates for daily soyfood use.[19] It should also be mentioned here that some concern has been raised with regard to allergies to soyfoods and soy-based infant formulas, as well as potential soybean antinutrients; however, soy protein is ranked 11th in allergenicity, with 0.5% of young children having an allergy to soy. The incidence of soy protein allergy among older children and adults is extremely rare.[19] As mentioned earlier, the FDA's approval of a health claim based on the association between consumption of soy protein and a reduced risk of coronary heart disease has significantly increased the demand for soy ingredients by the food industry. The government-approved health claim adds legitimacy to soy protein products, has helped to increase the awareness of soyfoods, and has created an incentive for food processors to add soy protein to foods; consequently, the number of soyfood products available has increased significantly.[20]

TABLE 4.5

Effects of Soy Flour on Bakery Applications

Absorption facilitates greater water incorporation.
Improves dough handling.
Improves machineability of cookie dough.
Improves moisture retention during baking.
Improves cake tenderness, crumb structure, and texture.
Enhances rate of crust color development.
Retards fat adsorption by donuts.
Prolongs freshness and storage stability.
Bleaches crumb color to produce white bread.
Improves nutritional quality.
Functions much like nonfat dry milk in bread and rolls, at about half the cost.

Source: Kinselle, J.E., *J. AOCS*, 56, 242, 1979. With permission.

Applications of Soy Ingredients in Baking

Soyfoods may be divided into four classes: soy ingredients, traditional soy-foods, second-generation soyfoods, and foods where soy is used as a functional ingredient. Soy ingredients include raw (or unprocessed) soybeans, soy flours (defatted and full-fat), soy concentrates, soy isolates, texturized vegetable soy protein, and hydrolyzed soy protein. Foods in which soy is used as a functional ingredient include baked goods to which soy flour is added.[21] Soy ingredients having unique nutritional and functional properties have found wide application in bakery products. The effects of soy ingredients in bakery products are summarized in Table 4.5, and bakery food applications of soy ingredients are provided in Table 4.6.

Nutritional Enhancement of Cereals

In recognition of the many advantages of soy flour as a protein supplement, considerable effort has gone into blending soy flour with cereals. These mixtures, when suitably fortified with vitamins and minerals, have great potential for feeding people of all ages in developing areas of the world. In addition to their significant levels of high-quality protein, the calcium and phosphorus levels of soy ingredients are also considerably higher than those of any other cereal grain, and soy flour is an excellent source of available iron. Furthermore, soy flour is considerably richer in vitamins than unenriched patent wheat flour and somewhat richer than enriched white flour.[13]

During the 1950s and 1960s, nutritionists sought to increase the adequacy and amount of dietary protein in developing countries. Soy-fortified grain products were seen as one way to accomplish this, and much work was done on incorporating soy into bread and pasta. Soy protein is relatively rich in

TABLE 4.6

Bakery Food Applications for Soy Ingredients

Ingredient	White Bread and Rolls	Specialty Breads and Rolls	Cakes	Cake Donuts	Yeast Donuts	Sweet Goods	Cookies and Biscuits
Defatted soy flour	X	X	X	X	X	X	X
Enzyme-active soy flour	X						
Low-fat soy flour			X	X		X	
High-fat soy flour			X	X		X	
Full-fat soy flour			X	X		X	
Lecithinated soy flour			X	X			X
Soy grits		X					
Soy concentrates		X					
Soy isolates		X	X	X	X		
Soy fiber		X					

Source: Dubois, D.K., *Soy Products in Bakery Foods*, Tech. Bull., Vol. II, Issue 9, American Institute of Baking, Manhattan, KS, 1980. With permission.

lysine but poor in methionine. Wheat protein, on the other hand, is poor in lysine but rich in cysteine (which the body can convert to methionine). Thus, the combination of the two protein sources creates a better balance of these two essential amino acids, which the human body cannot synthesize and must be obtained from our diet.[22,23]

A 1979 study found the protein efficiency ratio (PER) of gluten to be 0.7 and soy protein 1.3, but the PER for an 88:12 blend of wheat and soy flours was 2.0. Because of its low PER value, soy protein was long considered to be a second-class citizen in terms of protein value.[19] Since that study, however, nutritionists have turned to a different way of measuring protein nutritional adequacy, the protein digestibility corrected amino acid score (PDCAAS). In this test, casein (the protein in milk) is given the top score of 1.0; wheat gluten comes in at 0.25, and soy protein varies from 0.90 to 0.95, depending on its form. The protein in the 88:12 blend scores 1.0; thus, the improved nutrition achieved by combining the protein sources is apparent.[22,23]

A bakery-related potential use for soy flour in combination with cereals is in the production of the so-called "composite flours." These are mixtures of flours, starches, and other ingredients that are intended to replace wheat flour, totally or partially, in bakery products. Extensive research projects aimed at the development of such flours have been sponsored by international and national development agencies or programs such as the Food and Agriculture Organization (FAO), U.S. Department of Agriculture (USDA), U.S. Agency for International Development (USAID), and American Soybean Association (ASA) in the last 30 years or so. The main reason for developing composite flours is to relieve the burden of importing this commodity by countries where wheat is not grown.[5]

An important application of combining defatted soy flour and grits with cereals is the production of nutritionally balanced, all-purpose food blends to be distributed to undernourished populations or in food-shortage emergencies. The best known of these blends are corn–milk–soy (CMS), corn–soy (CS), and wheat–soy (WS). CMS contains 17.5% defatted soy flour, 15% nonfat dry milk solids, and about 60% corn. CS contains 22% soy flour and 71% corn. WS has 20% soy, 53% wheat bulgur, and 20% wheat protein concentrate. Another blend is Incaparina, which was developed to fight malnutrition in children.[5]

Cereal–soy blends or soy-fortified blended foods consist of a cereal grain (e.g., corn, wheat, sorghum) fortified with 10 to 30% soy flour or grits. The USAID PL 480 Food for Peace program distinguishes between blended food supplements (typically 15% defatted soy flour mixed with a cereal flour, then fortified with vitamins and minerals to be used in infant and child feeding) and fortified processed foods (typically 15% defatted soy grits or flour mixed with a cereal grit or flour). In Third World countries, the blend is typically made by extruding a blend of 70% dehulled corn and 30% dehulled whole (full-fat) soybeans. These low-cost, nutritious foods have come to be widely used in Third World countries since the mid-1960s.[24]

White Breads, Buns, and Rolls

Bread is a basic food product made of flour or meal, normally milled from wheat or rye but also from other cereals, that is mixed with water, salt, and other optional ingredients into a dough or batter fermented by yeast or leavened by other means, divided and formed into individual units of various weights and shapes, and baked.[3] Soy flours with minimal heat treatment (PDI of 80) are used at 0.5% to bleach flour, improve mixing tolerance, and impart flavor to the bread. Soy flours with a PDI of approximately 60 possess a milder flavor and are most commonly used at 1 to 2% in standard applications. These principles also apply to the production of buns and rolls.

In breadmaking, soy flour provides improved water-absorption and dough-handling properties, a tenderizing effect, body, and resiliency, as does nonfat dry milk (NFDM). Bread freshness is improved because the soy protein retains free moisture during the baking cycle. Also, soy protein products improve the crust color and toasting characteristics of bread.[4] Moisture absorption in bread increases by 4 to 5% for each 1% of isolated soy protein content.[12] In the United States, defatted soy flours are permitted in standardized bakery items at a maximum level of 3% (flour weight basis).[25] For enzyme-active soy flour, the FDA Standard of Identity permits a maximum of 0.5% (flour weight basis) in standardized bakery products.[26] No maximum has been established for other bakery foods for either defatted soy flour or enzyme-active soy flour.

It is common practice in breadmaking to add soybean lipoxygenase as part of the enzyme-active bread improver. It exhibits a gluten-strengthening effect and increases the mixing tolerance of a dough but also has an effect on staling rates and loaf color.[17,27] In general, when soy flour is held below 3% of the total flour added at the dough stage of the sponge-and-dough method, normal bread quality is obtained with the exception of a slight increase in absorption. Usually, for every kilogram of soy flour used in the formula, an additional 1 to 1.5 kg of water must be added. The addition of 3% soy flour to the sponge rather than dough stage adversely affects fermentation by weakening the gluten structure and reducing the gassing power, producing doughs that are sticky and difficult to make up and yielding bread with a coarse and uneven cell structure. For this reason, soy flour should be added at the dough stage when the sponge-and-dough method is used. With technologically improved soy products, such as soy protein concentrates or soy protein isolates, good-quality bread can be made using the straight dough method with soy product levels up to 5%; however, beyond that limit bread quality in terms of loaf volume, crumb color, and total score declines, particularly when even higher levels of oxidants and sodium stearoyl-2-lactylate (SSL) are used.[4,13] Nevertheless, Bing et al.[28] recently reported that it is the quality rather than the quantity of soy protein that influences bread quality. Likewise, Shogren et al.[29] also reported that an appealing, nutritious bread containing up to 30 to 40% soy flour can be prepared in an easy and economical manner using equipment available in home and institutional kitchens.

It is possible to reduce the production costs of some bakery products by adding soy ingredients to the formulas and making some adjustments in the production process. Porter and Skarra[30] evaluated the use of defatted soy flour as an extender in wheat breads, including sandwich bread, white pan bread, hamburger buns, and multigrain breads, due to its water-binding properties. In all cases, inclusion of soy flour in breads reduced material costs without adversely affecting the physical and sensory properties of the bread. They concluded that toasted soy flour can be used as a functional ingredient to improve the economics of bread production.

By using soy flour in conjunction with chemical oxidants it should be possible to moderate the undesirable effects of the latter. Thus, the combination of ascorbic acid and soy has been recommended as a general-purpose bread improver.[31] In the United Kingdom, where the use of benzoyl peroxide is no longer permitted and chlorine dioxide is virtually no longer used, the value of soy flour has grown. Likewise, a combination of ascorbic acid and soy flour, while no means an ideal replacement for potassium bromate, offers at least some improved tolerance and stability for long-processed doughs.[32]

A study conducted in 1974 by the American Institute of Baking (AIB) tested 14 soy protein products used to fortify bread. Results of this study showed that, while fortification with soy products affected bread quality somewhat adversely, high-protein soy products (concentrates and isolates) had more pronounced adverse effects even at a low (10%, flour weight basis) use level; however, most soy flours, especially full-fat and high-fat products, permitted fortification at levels of 15 to 20% and produced breads of acceptable overall quality that were high in protein.[33]

A notable achievement in the improvement of white bread was the development of a special bread by McCay et al.[34] at Cornell University. The bread is made using the Cornell Triple Rich Formula, which is so named because it contains soy flour, skim milk powder, and wheat germ. Recently, Zhang et al.[35] developed a highly acceptable soy-enriched bread ("soy bread") that contains sufficient soy proteins per serving to meet the FDA-approved health claim regarding soy and coronary heart disease.

Variety and Specialty Breads

Any bread other than the conventional, round-top pan bread made of only white flour can be categorized as a variety bread. It can include flours from other cereals, such as rye or whole wheat flour, or can be baked without the use of a confining pan directly on the hearth or on a baking sheet. Specialty bread is also any type of bread that differs from conventional round-top white bread in ingredients, shape, method of processing, crust, character, flavor, etc.[3] The protein content of ordinary white bread is 8 to 9%. Specialty breads can be made with 13 to 14% protein by incorporating soy proteins into a formula along with vital wheat gluten and, if necessary, a lipid emulsifier. Without an emulsifier, incorporating high levels of soy protein depresses loaf volume and gives poor crumb characteristics.[4] Special soy breads usually contain sufficient soy, in a mixture with wheat flour, to impart the characteristic flavor. The proportion of soy may be 15 to 25%. Defatted, enzyme-inactive soy flour (with a PDI of about 20) is darker and has a roasted flavor, thus it is the preferred material for breads in which a distinct soy flavor is required.[36,37] Soy protein isolates have been used for protein forti- fication of specialty breads because of their high protein content and bland- ness.[5] Lai et al.[38] reported that the addition of a high level of enzyme-active soy flour eliminated the need for the no-yeast sponge stage in the production of whole wheat bread with good volume.[38] In another application, using a mixture of 40% whole soy meal and 60% wheat flour and traditional fer- mentation technology utilized in the manufacture of soybean steamed bread (as compared to traditional steamed bread) produced golden bread with a rich aroma.[39]

Cakes and Pastry Products

Cake is a sweet, usually finely textured food product that is baked in various forms that differ in size and configuration. It generally contains such ingre- dients as soft wheat flour, milk or other liquids, sugar, eggs, chemical leav- eners, flavor extracts, and spices, as well as others that may or may not include shortening. Cake varieties cover a wide range and include pound cake, yellow and white layer cakes, cakes containing chocolate and cocoa products, sponge cakes, angelfood cake, fruit cakes, foam-type cakes, dough- nuts, and many others.[3] Soy protein products, including soy isolate–whey blends, have been used to produce commercially acceptable pound cakes, devil's food cakes, yellow layer cakes, and sponge cakes in which 50, 75, and 100% of NFDM has been replaced without impairing quality. At a 50% replacement level, aside from compensating for increased water absorption, no formula changes are necessary. With replacement levels at or above 75%, dextrose must be included in the total sugar used to improve color (except for devil's food cake). Leavening must also be increased to obtain the desired volume. The added cost of the leavening is offset by the increased yield of batter.[4] Soy protein products also help with the emulsification of fats and other ingredients. The resulting doughs are more uniform, smoother, more pliable, and also less sticky. The finished baked products have improved crust color, grain, texture, and symmetry and will stay fresher longer due to improved moisture retention.[4]

Defatted soy flour is used in cakes where specific water absorption and film-forming characteristics are desired. The typical usage level is 3 to 6% soy flour (flour weight basis). Benefits include a smoother batter having a more even distribution of air cells and a cake with a more even texture and a softer, more tender crumb.[6] Full-fat soy flour, alone or with refatted and lecithinated soy flours, is mainly used in the heavier types of cake batters, such as sponge cake and pound cake. It contributes to the richness of the cake while increasing the proportion of water that can be added to the mix. Due to their oil and phospholipids content, these flours exert egg and shortening sparing effects and act as emulsifiers. In these formulas, soy flour is used at a level of 3 to 5% (flour weight basis).[5] Utilization of defatted soy flour as a substitute for wheat flour in chiffon cakes was investigated, and it was concluded that chiffon cakes containing 15% defatted soy flour showed the highest sensory scores for overall acceptability.[40] Soy flour can be added to chemically baked products in a higher proportion than for yeasted products.[41] Short pastry items, such as pie crusts, fried pie crusts, and puff pastry, can be machined more easily and will retain freshness longer when lecithinated soy flour is used in the formula at levels of 2 to 4% (flour weight basis).[6]

Donuts

A donut (or doughnut) is a ring-shaped piece of sweet pastry that is cooked in deep frying fat to a rich brown color. Two basic types of doughnuts are produced: (1) cake doughnuts, made from chemically leavened cake batter, and (2) yeast-raised doughnuts, produced from a fermented sweet dough.[3] Cake donut quality is greatly improved by the addition of 2 to 4% (flour weight basis) defatted soy flour. The soy protein functions as a structure builder, providing a donut with an excellent star formation. Donuts containing soy protein absorb less fat during frying because the fat is prevented from penetrating into the interior. This may be due to heat denaturation of the protein on the donut surface, which produces a barrier to fat absorption. The result is a higher quality donut that is more economical due to lower frying oil use. Soy flour also gives donuts a good crust color, improved shape, higher moisture absorption and improved shelf life, and a texture with shortness or tenderness.[4,6,37] The effect of soy flour on fat absorption by cake donuts was studied by Martin and Davis.[42] They added soy flour to the cake donut formulation to reduce fat absorption during frying and concluded that the PDI is not a good indicator of soy functionality for this application; at all PDI levels above 50, the quantity of soy protein added is the best predictor of fat absorption control. Lecithinated soy flour is recommended for donut formulas containing minimal egg yolk levels because lecithin is the natural emulsifier in egg yolk that is such an important ingredient in cake donuts. It has been reported that approximately one half of the egg content of cake can be replaced with soy lecithinated flours.[4]

Cookies, Biscuits, and Sweet Pastry

A cookie is a small, cake-like product that is either flat or slightly raised. It usually, but not always, has a relatively low moisture content and is made from a dough or batter that is sufficiently viscous to permit the dough pieces to be baked on a flat surface. Cookies come in an infinite variety of shapes, sizes, composition, texture, tenderness, colors, and tastes. The term *cookie* is synonymous with *biscuit*, a term used in many countries.[3] In hard cookies, the use of 2 to 5% defatted soy flour improves machinability and produces a cookie having a crisp bite.[6] In cookies, the high water absorption properties of soy flour may restrict cookie spread and development of a typical top grain during baking; consequently, the baking performance and sensory qualities of these cookies could be adversely affected. Adding water to the formula could markedly improve the dough handling, spread characteristics, top grain, and sensory quality attributes.[43]

Low-fat soy flour, a major source of protein for dietary biscuits, may reduce the elasticity and increase the extensibility of doughs. Manley[44] advised caution with regard to claims that biscuits using 3 to 4% soy flour (wheat flour basis) have better appearance, better eating quality, and longer shelf life. The fat and lecithin (emulsifier) contents undoubtedly contribute to enhanced quality, but the water-binding characteristic of soy flour may not be ideal for biscuit dough (except when tightened consistency is desirable). Manley added that there is probably some value in considering soy flour as a replacement for eggs in a recipe, and for this purpose blends of soy flour and egg albumen have been prepared for use in wafer batter, for example.

In a pioneer work, Tsen et al.[45] concluded that soy flours (defatted and full-fat) and protein isolates improve the nutritive value of cookies by raising their protein contents and balancing their amino acids. Acceptable high-protein cookies can be prepared from wheat flour fortified with soy flour or protein isolate, particularly with the addition of sodium stearoyl-2-lactylate (SSL) or sodium stearyl fumarete (SSF).[45] Grover and Gurmukh[46] studied the cookie-making quality of five commercial defatted soy flours and reported that replacement of wheat flour by up to 15% soy flour is possible without adversely affecting the sensory characteristics of cookies.[46] Chocolate-chip cookies were produced by replacing 20% of the wheat flour with soy ingredients and were evaluated for color, flavor, texture, and overall appeal using a 9-point hedonic scale. Mean scores indicated that the panel, in general, liked the cookies.[47] While toasted soy grits give a nutty, toasted flavor to specialty cookies,[6] soy bran finds use in dietetic and healthfood biscuits.[44] In sweet goods, 2 to 4% defatted soy flour improves the water-holding capacity, sheeting characteristics, and finished product quality.[4]

Special Applications

In addition to their customary uses in baking, soy ingredients have other applications, including dietetic and diabetic breads. Although soy proteins do not contain gluten, they exhibit a strong binding power; thus, they find many applications in gluten-free bread production. A breadmaking mix for the manufacture of a dietetic bread is composed of rye flour (10%), wheat flour (30%), dry wheat gluten (20%), full-fat soy flour (10%), and very fine wheat bran (30%).[48] In a study conducted to assess the effect of starch-free bread on metabolic control in type 2 diabetes, Stilling et al.[50] used a starch-free bread containing soy protein.[49] In a study investigating the development of a premix for rice flour bread, the most acceptable bread was produced using 0.5% guar gum, 10% soy protein, and 105% water. Recently, Sanchez et al.[51] used response surface methodology to optimize gluten-free bread fortified with soy flour and suggested that acceptable bread could be prepared by adding 7.5% soy meal and 7.8% dried milk to the gluten-free formulation. Soy protein isolates proved quite functional in the production of gluten-free bread,[33] a product intended for use by gluten-intolerant individuals (celiacs). Good-quality gluten-free bread was made using wheat starch and up to 40% soy isolates by the American Institute of Baking.

Summary

Soy proteins are one of the most versatile and innovative food and ingredient systems. Because of their unique functional and nutritional properties, they have found wide application in bakery foods. The many different soy ingredients vary in their chemical composition and functional properties. Prior to selecting any soy ingredient, the desired characteristics of the food product should be carefully evaluated. With the incorporation of soy, appropriate adjustments in formulas and process parameters must be made to produce high-quality bakery products. Further improvements in raw materials and ingredient functionality coupled with innovative food-processing technologies will allow product developers to introduce a wide variety of new, more economical, naturally healthy, and attractive bakery products.

References

1. Jackel, S.S., Foreword, in *Wheat End Uses Around the World*, Faridi, H. and Faubion, J.M., Eds., American Association of Cereal Chemists, St. Paul, MN, 1995, p. v.
2. Seibel, W., Recent research progress in bread baking technology, in *Wheat Production, Properties, and Quality*, Bushuk, W. and Rasper, V.F., Eds., Blackie Academic, Glasgow, 1994, chap. 6.
3. Pyler, E.J., *Bakers Handbook*, Sosland Publishing, Kansas City, MO, 1994.
4. Endres, J.G., *Soy Protein Products: Characteristics, Nutritional Aspects, and Utilization*, AOAC Press, Champaign, IL, 2001, chaps. 5, 6.
5. Berk, Z., *Technology of Production of Edible Flours and Protein Products from Soybeans*, FAO Agricultural Services Bull. 97, Rome, 1992, chap. 4.
6. Dubois, D.K., *Soy Products in Bakery Foods*, Tech. Bull., Vol. II, Issue 9, American Institute of Baking, Manhattan, KS, 1980.
7. Schryver, T., Increasing health benefits using soy germ, *Cereal Foods World*, 47, 185, 2002.

8. Anon., Kerry unveils food industry's first whole-grain soy flour, *The Soy Daily*, February 25, 2005 (http://thesoydailyclub.com/Ingredients/kerry02252005.asp).

9. CAC, *Codex General Standard for Soy Protein Products*, Codex Standard 175-1989, Codex Alimentarius Commission, Rome.

10. Kinselle, J.E., Functional properties of soy proteins, *J. AOCS*, 56, 242, 1979.

11. Pringle, W., Full fat soy flour, *J. AOCS*, 51, 74A, 1974.

12. Hettiarachchy, N. and Kalapathy, U., Soybean protein products, in *Soybeans: Chemistry, Technology, and Utilization*, Liu, K., Ed., Chapman & Hall, New York, 1997, chap. 8.

13. Pyler, E.J., *Baking Science and Technology*, Sosland Publishing, Kansas City, MO, 1988, chap. 8.

14. Grosch, W., Redox system in dough, in *Chemistry and Physics of Baking*, Blanshard, J.M.V., Frazier, P.J., and Galliard, T., Eds., The Royal Society of Chemistry, London, 1985, chap. 12.

15. Morrion, W.R., Wheat lipids: structure and functionality, in *Wheat Production, Properties, and Quality*, Bushuk, W. and Rasper, V.F., Eds., Blackie Academic, Glasgow, 1994, chap. 9.

16. Cauvain, S. and Young, L., *Baking Problems Solved*, Woodhead Publishing, Cambridge, U.K., 2001, chap. 4.

17. Mathewson, P.R., *Enzymes*, American Association of Cereal Chemists, St. Paul, MN, 1998, chap. 5.

18. Wieser, H., The use of redox agents, in *Bread Making: Improving Quality*, Cauvain, S.P., Ed., Woodhead Publishing, Cambridge, U.K., 2003, chap. 20.

19. Anderson, J.W., Smith, B.M., Moore, K.A., and Hanna, T.J., Soyfoods and health promotion, in *Vegetables, Fruits, and Herbs in Health Promotion*, Watson, R.R., Ed., CRC Press, Boca Raton, FL, 2001, chap. 9.

20. Jordan, J., The market for soy proteins and textured products, in *Texturized Vegetable Protein Manual*, Riaz, M.N. and Baron, M.E., Eds., Texas A&M University, College Station, 2004.

21. Hendrich, S. and Murphy, P.A., Isoflavones: source and metabolism, in *Handbook of Nutraceuticals and Functional Foods*, Wildman, R.E.C., Ed., CRC Press, Boca Raton, FL, 2001, chap. 4.

22. Stauffer, C.E., More concentrated sources of soy protein in baking examined, *Milling & Baking News*, January 23, 2001.

23. Stauffer, C.E., Beneficial soy, *Baking Snack*, 26, 51, 2004.

24. Shurtleff, W. and Aoyagi, A., History of soy flour, grits, flakes, and cereal-soy blends, *History of Soybeans and Soyfoods: 1100 B.C. to the 1980s*, unpublished manuscript, Soyfoods Center, Lafayette, CA, 2004.

25. CFR, *Food and Drugs*, CFR 21, 136.110(c)(11), U.S. Code of Federal Regulations, 2004.

26. CFR, *Food and Drugs*, CFR 21, 136.110(c)(12), U.S. Code of Federal Regulations, 2004.

27. Casey, R., Lipoxygenase and breadmaking, in *The First European Symposium on Enzymes and Grain Processing*, Angelino, S.A.G.F., Hamer, R.J., van Hartingsveldt, W., Heidekamp, F., and van der Lugt, J.P., Eds., TNO Nutrition and Food Research Institute, Zeist, The Netherlands, 1997, p. 188.

28. Bing, Y., Renguo, R., and Zeny, Y., Study on application of soybean protein in breadmaking, *Food Science and Technology*, 4, 49, 2003.

29. Shogren, R.L., Mohamed, A.A., and Carriere, C.J., Sensory analysis of whole wheat/soy flour breads, *J. Food Sci.*, 68, 2141, 2003.

30. Porter, M.A. and Skarra, L.L., Reducing costs through the inclusion of soy flour in breads, *Cereal Foods World*, 44, 632, 1999.
31. Wood, P., Compound dough conditioners, in *The Master Bakers' Book of Breadmaking*, Brown, J., Ed., Turret Press, London, 1982, chap. 4.
32. Williams, T. and Pullen, G., Functional ingredients, in *Technology of Breadmaking*, Cauvain, S.P. and Young, L.S., Eds., Aspen Publishers, Gaithersburg, MD, 1999, chap. 3.
33. Ranhotra, G.S., *Commemorative Bulletin*, Tech. Bull., Vol. XXII, Issue 4, American Institute of Baking, Manhattan, KS, 2000.
34. Ensminger, M.E., Ensminger, A.H., Konlande, J.E., and Robson, J.R.K., *The Concise Encyclopedia Foods and Nutrition*, CRC Press, Boca Raton, FL, 1995, chap. 2.
35. Zhang, Y.C., Albrecht, D., Bomser, J., Schwartz, S.J., and Vodovotz, Y., Isoflavone profile and biological activity of soy bread, *J. Agric. Food Chem.*, 51, 7611, 2003.
36. Bread Research Institute of Australia (BRIA), *Australian Breadmaking Handbook*, Tafe Educational Books, Kensington, 1989, chap. 10.
37. Matz, S.A., *Ingredients for Bakers*, Pan-Tech International, McAllen, TX, 1987, chap. 2.
38. Lai, C.S., Davis, A.B., and Hoseney, R.C., Production of whole wheat bread with good loaf volume, *Cereal Chem.*, 66, 224, 1989.
39. Suchun, L. and Wuming, H., Preparation of soybean steamed bread, *Food Sci. Technol.*, 1, 20, 2003.
40. Punbusayaku, N., Utilization of defatted soy flour for partial replacing of wheat flour in chiffon cake, *Food*, 29, 180, 1999.
41. Amendola, J. and Rees, N., *Understanding Baking*, 3rd ed., John Wiley & Sons, New York, 2003, chap. 1.
42. Martin, M.L. and Davis, A.B., Effect of soy flour on fat absorption by cake donuts, *Cereal Chem.*, 63, 252, 1986.
43. McWatters, K.H., Cookie baking properties of defatted peanut, soybean, and field peas, *Cereal Chem.*, 55, 583, 1978.
44. Manley, D., *Technology of Biscuits, Crackers and Cookies*, 3rd ed., Woodhead Publishing, Cambridge, U.K., 2000, chap. 9.
45. Tsen, C.C., Peters, E.M., Schaffer, T., and Hoover, W.J., High protein cookies. I. Effect of soy fortification and surfactants, *Bakers' Digest*, 47, 34, 1973.
46. Grover, M. and Gurmukh, S., Evaluation of commercial defatted soy flours for cookie making: effect on physical and sensory characteristics, *J. Dairy, Foods Home Sci.*, 13, 91, 1994.
47. Chen, D.J., Weingartner, K., and Brewer, M.S., Consumer evaluation of soy ingredient-containing cookies, *J. Food Qual.*, 26, 219, 2003.
48. Riegel, M.L.A., Breadmaking Composition, Bread and a Process for Preparation of a Dietetic Bread Based on Rye and Wheat Flour, French Patent Application No. FR 2 753 344 A1, 1998.
49. Stilling, B., Mehlsen, J., Hamberg, O., Larsen, J.J., Gram, N.C., and Madsbad, S., Effect of starch-free bread on metabolic control in type 2 diabetes, *Lancet*, 352, 369, 1998.
50. Sodchit, C., Kongbangkerd, T., and Weeragul, K., Development of premix for rice flour bread using guar gum as a binder, *Food*, 33, 222, 2003.
51. Sanchez, H.D., Osella, C.A., and Torre, M.A., Use of response surface methodology to optimize gluten-free bread fortified with soy flour and dry milk, *Food Sci. Technol. Int.*, 10, 5, 2004.

5

Developing and Producing Protein-Enhanced Snacks and Cereals

Brad Strahm

CONTENTS

Introduction

A popular trend that has been developing over the last several years is protein-enhanced snack and cereal products. This trend has been driven by increased consumer interest in low-carbohydrate diets and in the health benefits of soy protein and other vegetable protein sources. Protein-enhanced products are typically designed to mimic products that are based primarily on carbohydrate-rich recipes — that is, products made from cereal grains or tubers or starches isolated from cereal grains or tubers. These products can be made by any number of methods that are typically used to make snack products, including extrusion. Due to the popularity of extrusion processing and its application to this category of products, protein-enhanced snack and cereal products made via the extrusion process are the focus of this chapter.

Characteristics of Protein-Enhanced Snack and Cereal Products

Protein-enhanced snack and cereal products are required by processors and consumers to have several important characteristics. These characteristics also represent areas where challenges exist in the development and production of protein-enhanced snack and cereal products. First, the texture of these products should be crispy or crunchy and certainly should not have a hard texture. Second, these products must have a good flavor, which is usually the most important consumer criterion. Third, they should be resistant to breakage during processing, which is primarily important to the processor.

Two Approaches to Protein-Enhanced Snack and Cereal Products

Protein-enhanced snacks are based on essentially two different extrusion-based approaches. The first approach is to base the production of protein snacks on the extrusion technology for texturized vegetable proteins. The second approach is based on using the extrusion technology for cereals and snacks. As discussed in an earlier chapter, extrusion is widely used to manufacture texturized vegetable protein products. Textured vegetable proteins are normally hydrated to achieve meat-like textures to mimic meats. With the amount of expansion and the associated bulk density typically used for texturized vegetable proteins, the texture of these products can be too hard and crunchy; however, these products can also be extruded with greater expansion and lighter density, resulting in a slightly lighter, less crunchy texture that would be suitable for eating as a snack product. Greater expansion is accomplished either by simply extruding under conditions that supply more energy to the product, resulting in a greater driving force for the expansion process, or by adding starch to the product. By adding starch, which expands better than protein, a greater degree of expansion can typically be accomplished. Products made utilizing this technology are also often flavored with savory flavors that are externally applied after the product is dried. The flavors will typically be adhered to the base product using vegetable oil, and the seasonings are either mixed with the oil prior to being applied or added in a separate stream to product that has been sprayed with the oil.

Breakfast cereals and snacks are products that are often extruded. They are typically based on milled fractions of the cereal grains. Corn is the most common cereal grain used for manufacturing extruded snacks. Corn, wheat, rice, and oats are commonly used for manufacturing extruded cereals, but other grains, such as barley and sorghum, can also be used. In addition to the cereal grains, breakfast cereals typically contain a significant amount of sugar, which is added for flavor and textural reasons. While some breakfast cereals are over 50% sugar as they are consumed, the extruded base will typically contain 6 to 8% sugar, and the remainder is added externally. The general category of breakfast cereals also includes extruded crisp rice, which is commonly used in snack bars, where the extruded crisp rice is combined with other elements such as syrups, chocolate, and nuts to create the snack bar. The crisp rice used in these bars is almost always extruded, and this is a major market where protein-enhanced crisp rice, often referred to a protein crisps or soy crisps, is being utilized in today's market.

It is desirable that breakfast cereals and crispies that are used in snack bars have a crispy texture that results in some crunch but is not too soft. When protein is added to these products, the expansion is typically reduced and a harder, crunchier texture results.

Developing Protein-Enhanced Snack and Cereal Products

When food scientists go through the process of developing protein-enhanced cereals and snacks, they generally are given the task of creating a product with a higher protein content that is similar to another product that is readily identified by consumers. For example, a product developer might be given the task of creating a protein-enhanced Cheetos® or a protein-enhanced Froot Loop®. Given a task such as this, the natural course for the product developer to take would be to first become familiar with the reference product, including what raw materials it is made of and how it is made. The next natural step might be to simply use the same process and replace all or part of one of the high-carbohydrate ingredients with a high-protein ingredient in order to produce a product with a higher overall protein content. What should we expect the results of this approach to be? In order to better understand what the expected results should be, we first need to better understand what it is that gives the reference product, the one for which we are now trying to create a high-protein version, its identity. In a simplified sense, the identity of the reference product is a result of two main factors: what it is made of (the ingredients) and how it is made (the process). If we significantly change at least one of these two key factors, then we should expect to get a product with a different identity!

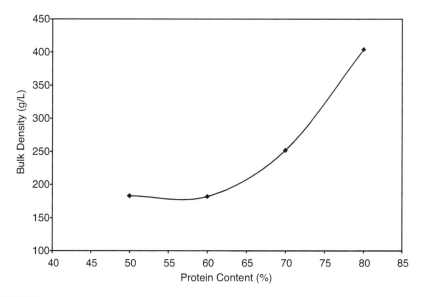

FIGURE 5.1
Effect of added protein on the expansion characteristics of extruded products.

Anticipated Challenges in Product Development

Now, revisiting the approach that a product developer might be likely to take (that is, changing the ingredients to formulate to a higher protein content and using the same process used to create the reference process), the product developer should not reasonably expect to get a product with the same identity. That is, the developer should not expect to end up with a product with the same texture, density, crunchiness, color, etc. Several problem areas are typically encountered in the development of protein-enhanced snacks and cereals: expansion and texture, shaping, mouthfeel, flavor, and raw material costs. These problems are all basically related to the inherent differences between primarily starch materials in the cereal grains and the proteins that are used to boost the protein content in protein-enhanced products.

Reduced Expansion

Adding protein to starch materials typically results in less expansion, higher density, and harder texture. Figure 5.1 shows this effect for a protein crisp with varying levels of protein content. In this simple study, the product was expanded as much as possible using a twin-screw extruder. The individual recipes were combinations of long grain rice flour and a protein source that was either soy protein concentrate or soy protein isolate.

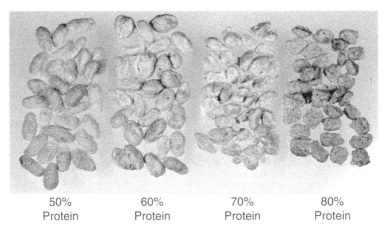

| 50% Protein | 60% Protein | 70% Protein | 80% Protein |

FIGURE 5.2
Effect of added protein on the shape of extruded crispies.

Difficulty in Shaping

Protein-enhanced products are typically more difficult to form into an eye-pleasing shape at the extruder discharge. This is related to the differences in flow characteristics between starch and protein. Starch flows easily and expands uniformly, while protein tends to flow and expand in a more non-uniform manner. The resulting shapes therefore appear deformed. This is demonstrated by the products shown in the photograph in Figure 5.2. Note that, as more protein is added, the product pieces become more irregular in shape, revealing greater difficulty in shaping the product.

Raw Material Costs

While product function, texture, and appearance are often the primary focus of a product developer, the cost of the materials that are combined to create the product is also an important consideration. In the case of developing high-protein snacks and cereals, costs can be a very important consideration because the costs of the protein-enhancing ingredients are often much higher than those for the high-carbohydrate ingredients typically used to create this kind of product. Proteins from a variety of sources can be used, and they can be either animal or plant based. These protein sources include animal proteins such as egg, milk, meat, and gelatin, as well as plant proteins such as soybean, pea, peanut, canola, sesame, and wheat gluten. Table 5.1 shows the relative costs of several ingredients that a product developer might use in developing a protein-enhanced product compared to corn and wheat flour. Note that many of these ingredients are quite expensive, up to over 20 times more expensive than the high-carbohydrate materials they might replace in a protein-enhanced formulation. Of course, the materials that are used as

TABLE 5.1

Relative Costs of Raw Materials for
Extruded Products

Raw Material	Cost Ratio
Yellow corn meal	1
Wheat flour	1
Soy flour	2
Vital wheat gluten	3
Whey concentrate (34%)	5
Soy isolate	11
Wheat protein isolate	19
Powdered egg whites	21

TABLE 5.2

Relative Costs of High-Protein
Ingredients

Raw Material	Cost Ratio
Soy flour	1
Vital wheat gluten	1
Soy isolate	3
Whey concentrate (34%)	4
Whey concentrate (80%)	5
Wheat protein isolate	6
Powdered egg whites	7

protein sources have varying levels of protein. Because a formulator's likely objective is to formulate to a target protein level, it may be necessary to use more of a low-protein ingredient and less of a high-protein ingredient to reach the targeted protein level. In Table 5.2, several protein ingredients are listed along with their relative cost per unit of protein. The ingredients that tend to cost more per unit of protein are typically those that are more refined and have higher protein levels.

Poor Flavor

As already discussed, when a high-protein version of a high-carbohydrate product is developed, it changes the identity of that product. Many high-carbohydrate products have a characteristic flavor that is associated with the materials and processes used to create them. When the ingredients are changed to high-protein ingredients that have their own characteristic flavor, the flavor is changed. In addition, through the thermal process used to create the products, these ingredients react differently than the high-carbohydrate ingredients do, such that different flavor profiles are created. A distinctly

"chalky" mouthfeel or flavor is commonly associated with high-protein snack
and cereal products. This chalky mouthfeel is likely related to the fact that
thermally processed proteins are typically insoluble in water. By contrast, the
cooked and sheared starch in high-carbohydrate products, which do not have
this chalky mouthfeel, are typically highly soluble in water. Thus, when eating
a starchy snack, the starch readily solublizes in saliva and is flushed out of
the mouth. The insoluble proteins in high-protein snacks does not readily
solubilize in saliva and therefore remains in the mouth longer, resulting in
the chalky sensation. This, in addition to the textural problems, is likely the
greatest challenge in developing high-protein products.

Anticipated Challenges in Production

After developing a product that is designed to meet the desires of the target
population of consumers, the next step is typically to move the product from
the pilot plant where it was developed to the production processing plant.
In doing so, several challenges may be encountered, including having the
proper equipment available in the production plant, the cost of the raw
materials, reduced yield in the processing plant, variations in the raw materials, and shaping of the product. Some of these problems are the same as
those encountered during the product development process, and they
remain challenges during production.

Inappropriate Equipment

As we discussed earlier, high-protein products are often developed using
equipment similar to that used to make the high-carbohydrate reference
product and by simply changing the ingredients to a blend with a higher
protein level; however, in many cases, the optimum equipment configuration
required for making the product may not be the same as that used for the
high-carbohydrate reference product. If products are developed in a typical
pilot plant where the equipment is typically configured for a high level of
flexibility, the configuration of the equipment in the pilot plant can be optimized to produce the high-protein product; however, the same degree of
flexibility is often not built into the production facility itself. For this reason,
it is best to have prior knowledge of the target production facility and
develop the product using tools that are available in that facility. Unfortunately, product development is often conducted with no particular target
production facility in mind.

In the case of extruded products, an example of an equipment configuration problem that may be encountered is the preconditioning step prior to
the extruder. The photograph in Figure 5.3 shows the advantage in shape
and color uniformity that preconditioning can provide when making

FIGURE 5.3
The advantages of preconditioning: with (left) and without (right).

extruded protein crisps. If this advantage is discovered during the develop-ment phase in a pilot plant where a preconditioner is available and a pre-conditioner is not available in the production facility, then the production facility may encounter difficulty in producing the more uniform shape and color product.

Another example where improper equipment may be encountered in a production facility is the case where the extrusion feeding device is designed to handle a uniformly granular material such as corn meal, which is a material having a very uniform particle size distribution. Materials such as this are sometimes delivered into the extruder using a vibratory feeding device that will not feed small particle size, or floury, materials uniformly. When a protein-enhanced product is developed, the particle size distribution is likely to contain much smaller particles that are not compatible with the vibratory feeder typically used for delivering corn meal into the extruder.

Raw Material Costs

The impact of raw material costs on the product development process has already been discussed. This impact, however, is magnified when the pro-duction stage is reached. While the raw material used to make an extruded product is typically the most expensive cost input, this effect is magnified when the recipe is changed to expensive protein raw materials. In this situ-ation, the cost of any activity that results in the loss or wasting of raw materials is magnified significantly.

Plant Yield

One of the critical performance factors for any production plant is the yield of the plant. Plant yield can be simply expressed by comparing the quantity of product produced to the quantity of raw materials consumed. A number of things can reduce plant yield, including the production of off-specification product, fines, startup waste, shutdown waste, spillage, and producing finished product at a moisture content that is lower than that of the raw materials. An average estimate of plant yield is about 85%. For high-protein products where the production is more difficult and sensitive, several of these factors play an important role. Products that do not meet the required specifications are produced more often due to the difficulty in achieving expansion and controlling shape. Due to problems with blistering or with good cutting, fines can also be a significant problem. While plant yield is always important to a profitable operation, it is especially important to the production of high-protein products due to the increased raw material costs, as already discussed.

Variations in Raw Materials

The raw materials that are used to make any product and the characteristics of those raw materials are always the most important factor in the successful production of an extruded product. Because of all of the production difficulties already discussed, many high-protein products are produced in a processing window that pushes the limits and capabilities of the equipment and process being employed. Because of this, if the characteristics of the raw materials change, it may no longer be possible to produce the product within the set specifications. These raw material changes may be related to a number of factors. The after-ripening factor often shows up during the change from one crop year to another.

It is well known that the processing characteristics of grains can change when they are stored versus newly harvested grains. In addition, many of the protein materials that are used in the production of high-protein products are highly processed prior to being used to manufacture the final product. When this is the case, each processing step introduces an opportunity for sometime very subtle changes to be induced in the product, resulting in large changes in how the materials perform in the production line. Lot-to-lot raw material variations that do not show up in the normal specifications used by the raw material manufacturer have been a problem that has plagued the high-protein snack industry. A number of analytical tools have been used to try to better understand and therefore adjust for these changes, including the Rapid Visco-Analyzer® (RVA; Newport Scientific) and the Phase Transition Analyzer® (PTA; Wenger Manufacturing), as well as a number of more conventional techniques.

Summary

Over the past few years, interest in high-protein versions of normally starchy snacks has increased. This trend has been driven by dietary trends as well as health recommendations by the U.S. federal government. While the high-protein/low-carbohydrate dietary trend may not last in it most vigorous form, certainly an element of this approach to dietary health will survive. For this reason, food processors will continue to be interested in products that fall into this category for the foreseeable future. While some good products that fall into this category have been produced, a number of challenges still exist with regard to developing and producing high-protein snacks and cereals. As with most food products, taste is the most important factor, and it continues to be a major challenge in the development of these products.

6

Soy in Pasta and Noodles

Wesley Twombly and Frank A. Manthey

CONTENTS

Introduction

The simplicity of pasta and noodles makes them a suitable carrier for a variety of nontraditional ingredients. Marconi and Carcea[1] summarized reasons to add nontraditional ingredients to pasta products. Reasons they cited that are applicable to the addition of soy products to pasta include:

- *Nutritional improvement* — Pasta was among the first foods in the United States to be vitamin fortified. It is a good carrier for other ingredients, as well, including high-protein flours such as soy products.

- *Use of byproducts* — Typically, byproducts are used to improve a nutritional component of the pasta, such as adding in dietary fiber.

- *Health benefits* — The examples provided in the article focus on dietary fiber, but adding the isoflavones from soy to pasta is an example of potential fortification for health benefits.

Soy contains nutritional and healthful ingredients, such as dietary fiber, protein, and isoflavonoids. Soybeans contain approximately 18% lipids, 42% protein, 14% neutral detergent fiber, 15% total nonstructural carbohydrates, and 13.5% total sugars. The environmental conditions under which soybeans are grown have a great impact on their chemical composition and nutrient quality.[2,3] Soy lipid is rich in polyunsaturated fatty acids (PUFAs) and lethicin. Soybean protein is rich in the essential amino acids arginine, leucine, lysine, phenylalanine, and valine. Total sugars consist primarily of sucrose, stachyose, and uronic acid. Soybean is rich in isoflavonoids, which have been reported to protect against health problems associated with menopause, cancer, and cardiovascular disease.[4]

Over the past few years, low-carbohydrate diets seem to have been the primary reason for incorporating soy into pasta products. Pasta is a product that consumers do seem to miss when it is removed from their diets, so it is reasonable to try to develop pasta products that reflect changes in dieting trends. In addition to low-carbohydrate diets are managed-carbohydrate diets, where the intake of calories is controlled to include some percentage of each dietary component: carbohydrate, protein, and fat.

A potential market that seems to have been overlooked the past few years is the functional food market, a market where a U.S. Food and Drug Administration (FDA)-approved health claim can be attached to a product. In the case of soy protein, the claim is a "heart healthy" statement. To receive the heart health claim, it is necessary for a pasta product to provide 6.25 g of soy protein per serving of pasta (21 CFR 101.82).

General Pasta and Noodle Processing

Traditional Pasta

Traditional pasta is made only from durum wheat semolina and water. When the supply of semolina is limited, common wheat flour or farina is used as a replacement for durum wheat semolina. Traditional pasta relies on the storage proteins in wheat endosperm to form a protein (gluten) matrix, which gives pasta its mechanical strength and cooked quality. Traditional pasta processing can be divided into three stages: mixing, kneading, and extruding.

Mixing

During the mixing stage, water is sprayed onto the dry semolina at a rate that increases semolina moisture content from about 14% to 30–32%. Water sprayed onto the semolina has a slightly elevated temperature (35 to 40°C) which aids water absorption into the semolina particles. The mixing chamber typically has rotating paddles that promote uniform wetting of the semolina surface, prevent wet semolina granules from clumping together, and move the semolina toward the extrusion barrel. Wetted semolina remains in the mixer long enough for moisture to be absorbed into the semolina granule. Retention time in the mixing chamber typically ranges from 10 to 20 minutes. Most dry pasta manufacturers apply a partial vacuum (–63 to –80 kPa) to the mixing chamber.[5] This vacuum promotes hydration by eliminating the surface tension associated with air and reduces pigment oxidation by lipoxygenase enzymes.[6] The vacuum also prevents air from being trapped inside the extruded pasta; such trapped air can create focal points for stress during drying at high temperatures.

Pasta Extruder

The extruder consists of a screw and a barrel. The extrusion barrel has a water jacket as a means of controlling the dough temperature. Warm water circulates quickly within the water jacket to remove excess heat generated by the friction between the dough and the extrusion barrel. The dough temperature is kept between 45 and 50°C, as a warm dough temperature reduces the viscosity of the dough. Temperatures above 50°C must be avoided in order to prevent the gluten proteins from denaturing. The inside bore of the extrusion barrel is grooved with longitudinal grooves which prevent the dough from spinning at the wall. The extrusion screw has a length-to-diameter ratio of from 6:1 to 9:1, which results in low mechanical energy/unit throughput.[5] Pasta screws typically are deep flighted and have a constant flight height and uniform pitch the entire length. The screw rotates at 15 to 30 rpm.

Kneading and Extruding

Kneading and subsequent dough development occur inside the extrusion barrel. The hydrated semolina becomes compacted as it moves into the extrusion barrel. After compaction, the dough mass moves forward due to friction between the dough and the extrusion barrel. Restricted flow, due to the die and the kneading plates, causes back-pressure in the extrusion barrel. The forward flow of dough and the backward pressure cause kneading of the dough, which develops a protein matrix that entraps starch granules. The protein matrix develops in the direction of extrusion. The developed dough is then extruded through the die that is located at the end of the extrusion barrel. The shape of the pasta is determined by the shape of the die orifice.

Preservation

Pasta can be dried or packaged fresh without drying, but most pasta is dried before packaging. The moisture content of pasta extruded from the die is about 29 to 31%. The objective of drying is to reduce the moisture content to less than 13% without creating any undue stresses within the complex protein and starch structure. Drying occurs by regulating airflow, air temperature, and relative humidity.[7] Pasta is a product that is very difficult to dry. Pasta has a relatively smooth surface, which limits the capillaries available to convey moisture to the surface of the product, and it has a very low diffusivity. The traditional drying of pasta in a Mediterranean environment could require in excess of 40 hours for a product such as spaghetti. As the technology improved, drying cabinets and drying conveyors were created. Over time, three different types of drying developed: conventional drying (60°C maximum temperature during the drying cycle), high-temperature drying (~80°C maximum temperature during the drying cycle), and ultra-high-temperature drying (temperatures exceeding those of high-temperature drying).[8] As the maximum temperature during drying increases, the drying time decreases. An ultra-high-temperature drying system can dry spaghetti within 5 hours. Drying the pasta too fast, however, will result in checking (cracking). If the pasta is held under the incorrect time, temperature, and humidity environment, Maillard browning will occur.[9,10] Presumably, adding soy protein would result in added free amino groups, which would make the Maillard browning reaction more likely to occur.

Traditional Wheat Noodles

Wheat noodles are made from flour of soft or hard common wheat or durum wheat and salt water. Salt is added for flavor, to strengthen the gluten network, to inactivate enzymes in the dough, and to inhibit dough fermentation during processing. Wheat noodles vary in their texture and appearance.[11] For example, Japanese udon noodles made from soft white wheat with low protein (8 to 9.5%) are very white and have a soft cooked texture. In contrast, Chinese alkaline noodles made from hard wheat with a high protein content (11 to 12.5%) are yellow and have a chewy cooked texture. Like pasta, wheat noodles rely on storage proteins in wheat endosperm to form a protein (gluten) matrix, which gives noodles their mechanical strength and cooked quality. Processing of wheat noodles can be divided into five stages: mixing, dough sheeting, combining of two sheets, reducing, and cutting.

Mixing

The hydrating and blending of ingredients for noodle processing are similar to those described earlier for pasta processing. Noodle ingredients are generally hydrated with warm water to a water content of 32 to 40%, somewhat higher than that of hydrated semolina. The mixing time is about 5 to 10 minutes. After mixing, the wetted ingredients are rested for 20 to 40 minutes before combining. Resting aids even water penetration into the dough, resulting in a smoother and less streaky dough after sheeting. Proper dough (protein matrix) formation requires complete hydration of flour particles. Incomplete hydration will give the noodles a streaky appearance, corresponding to areas of incomplete protein matrix development. Commercial production dough is rested in a receiving container while being stirred slowly.[12]

Combining

The hydrated material is divided into two portions, and each portion is passed through a pair of combining rolls to form a thick noodle sheet. The two sheets are then combined and passed through a second set of rolls to form a single sheet. The dough is developed by the pressure generated as the dough passes between the rolls. The combined dough sheet is placed into a temperature- and humidity-controlled chamber for up to 1 hour.[13] This stage of dough resting allows the dough to lose some of the stresses that built up during dough development. Hou[12] listed four functions of resting the dough: (1) to help distribute moisture more evenly, (2) to enhance disulfide bond formation, (3) to form bonds between gluten and lipids, and (4) to relax the gluten for easy reduction in subsequent sheeting operations.

Rolling

Three to five pairs of rolls, each having a smaller diameter with decreasing roll gaps, are used for the sheeting process. A slow reduction in thickness is necessary in order to prevent damage to the gluten network. Dough reduction of 30% is preferred so the gluten can maintain its intact structure.[12] Repeated sheeting can increase the density of the noodles by pressing out gas. The final roll gap must be narrower than the desired thickness, because the noodle will expand slightly after passing through the rolls.

Cutting

The sheet is cut into strands of the desired width by passing the dough sheet through a pair of cutting rolls.

Preservation

After cutting, wheat noodles are consumed fresh within 24 hours of manufacturing, parboiled, steamed, or hung on rods similar to pasta and dried. According to Nagao,[13] the first step in drying noodles is the predrying stage, where moisture is reduced from 40 to 27%. This is typically done at a low temperature (15°C) and with dry air. The predrying stage lasts 0.5 to 1.5 hours. The second step is the sweating stage, where internal moisture diffuses to the surface. Salt added to the dough affects moisture diffusion. Salt helps in the diffusion of moisture from the inside to the drier outer surfaces. In the sweating stage, the air is often heated to 40°C, and the relative humidity is maintained at 70 to 75%. The sweating stage lasts 3 to 5 hours. In the third stage, the noodles are cooled to ambient temperature and humidity.

Processing Pasta and Noodles Containing Soy

Two strategies are used to manufacture pasta or noodles containing nontraditional ingredients. One is to rely primarily on the gluten matrix formed by wheat proteins for the structure and strength of the product. The other is to rely on gelatinized starch for the structure and strength.

Strategy One: Structure from Gluten Matrix

Gluten formation requires that the wheat storage proteins be hydrated and kneaded. Non-wheat ingredients can physically interfere with dough development by causing discontinuity within the gluten matrix.[14–17] The presence of non-wheat ingredients can affect dough formation and strength. When adding soy into pasta, semolina is displaced from the formula. This means that both gluten proteins and starch are removed from the formula. Gluten is the primary structure component for traditional pasta, so having a formula with a reduced gluten content often reduces the mechanical strength and cooking quality of the pasta.[15–17] Some of the wheat gluten can be displaced with minimal impact on pasta properties. A common rule of thumb seems to be that any addition of a non-wheat ingredient into pasta up to about 10% is acceptable; beyond 10%, the quality of the pasta degrades relatively quickly due to the gluten matrix being too greatly diluted.

None of the sources of soy protein is pure protein, so the total added ingredients are greater than just the amount of soy protein added. If a health claim for the added soy is desired, the level of soy protein isolates added must exceed the level of non-wheat ingredients that can be added without effect on the pasta. A soy protein isolate at 90% protein, on a dry weight basis, would require the addition of 6.94 g of isolate per serving (assuming 5% moisture in the isolate and 12% moisture for the finished pasta) to achieve the health claim. This means that approximately 14% of the weight of the pasta must be soy protein to satisfy the health claim in a 56-g serving. When making pasta that contains non-wheat ingredients at a level capable of impacting the quality of the pasta, a structure builder is needed to maintain the integrity of the pasta. To maintain the structure of the pasta, a structure builder must be added into the formula. Some structure builders and the levels at which they may have to be added include:

- *Vital wheat gluten*, which should be approximately 14% of the weight of the ingredients in the formula that are not semolina or durum flour. This approximation assumes that the level of gluten determines the structure of the product.
- *Egg albumin*, which should be up to 2% of the weight of the total formula. This estimation is based on the level of egg albumin used to make egg noodles.
- *Gelatinized starch*, which should be 30% or more of the weight of the ingredients in the formula that are not semolina or durum flour. This appears to be a workable minimum level, and increased levels (up to 100%) can typically be used with good results.

These numbers are only approximations, and finding the correct structure builder and level for any given formula will require experimentation. As most of the products commercially available seem to be formed by extrusion, the discussion in this chapter focuses on extruded pasta; however, many of the basic concepts will transfer well to sheeted products.

Issues with Blending Soy with Semolina/Flour

The blending of soy ingredients with semolina can raise a number of issues. Semolina is a particle with much larger particle size and bulk density than soy ingredients. Differences in particle size and density are contributing factors to segregation in blends. One way to minimize segregation is to try to more closely match the density and particle size of the durum with the soy. In addition to semolina, durum flour is readily available and has a particle size much closer to soy products. It must be noted that pasta extruders are choke fed, so they are actually volumetric devices. Any decrease in density of the material fed in will result in a proportional decrease in the throughput of the pasta extruder. Alternatively, it is at least conceptually possible to select soy products that have been agglomerated to have a particle size more similar to semolina, but it does not appear as though any readily available agglomerated soy products are on the market. Another possible solution is to have each of the ingredients metered into the mixing chamber separately.

Issues with Soy Ingredients in the Mixing Chamber

Typically, a rotary airlock is located directly before the mixing chamber on a pasta system. Some soy ingredients have a tendency to bridge, which can cause processing problems at either the feeders or the airlock. Some solutions that can be tried include the use of flow agents, live-bottom bins, and vibrators or shakers. Some soy ingredients can build up in airlocks (or other locations), so inspection of the equipment is necessary until the ingredient has some processing history behind it to confirm that buildup will not be a problem. Buildup can result in a loss of system capacity due to caked-on ingredients occupying some of the airlock volume of the system. Microbial growth could occur if the ingredients that build up in the rotary airlock develop a high water activity by absorbing moisture from the mixer environment.

Soy ingredients tend to be very fine particles with a low density, so the ingredients can become airborne very easily; thus, a noticeable amount of soy can be drawn into the vacuum system. The soy ingredients that find their way into the vacuum system can be deposited on the walls of the piping, at any filter, or at a point in the vacuum system where particles are likely to touch and adhere to the wall of the plumbing. In extreme circumstances, it is possible for the piping to be completely plugged due to this buildup.

The rate of moisture absorption in the ingredients may be an issue that requires attention. If the wheat-based ingredients (semolina, durum flour, or vital wheat gluten) absorb water more quickly than the soy ingredients, the result is likely to be a pasta that has regions that are very clearly wheat based and areas that are very clearly non-wheat based. This type of defect is likely to result in a white or dry appearance. It may cause additional cooking loss, but the product is likely to maintain its shape through the entire process. The other extreme is when the wheat-based ingredients are moisture starved. If this occurs, the gluten matrix will not develop, and the pasta will tend to fall apart at the die, in the dryer, or during cooking. This issue can be addressed to some extent by choosing ingredients with similar hydration times, which may be fine-tuned slightly by modifying the particle size of the ingredients (smaller particles will absorb moisture more quickly). At least in concept, it would be possible to partially or fully hydrate the ingredients prior to mixing them together, which would eliminate the issue of moisture competition.

Some soy ingredients may also require a long period of time to fully hydrate. This hydration time may exceed the time in the mixing chamber and pasta press. If the soy ingredients reach full hydration sometime after exiting the die, very visible voids will appear in the pasta as the moisture redistributes. Partially hydrating the soy ingredients prior to introduction to the mixing chamber may be a way to limit this issue. Presumably, the soy ingredients could be hydrated to increase their moisture content without causing a microbial growth problem. To the best of the authors' knowledge, this has not been tried on any trial or production scales.

Because the system is under vacuum, the formulated moisture and the actual dough moisture will not match. In a system with a long mixing time and using semolina, 2% of the formula moisture may be lost to the vacuum (e.g., formulated moisture of 33% but an actual extrusion moisture of 31%). Formulas with soy tend to experience an even higher moisture loss to the vacuum. It is possible to run pasta systems without a vacuum to minimize issues with moisture loss and to eliminate problems with soy getting into the vacuum system. Additional problems may occur, however. The primary problem would be additional buildup in the airlock at the mixing chamber. The pressure change as the airlock opens to the mixing chamber seems to help force product out. Without the vacuum, more rapid choking of the airlock may occur.

The hydrated, mixed ingredients exit the mixing chamber by being conveyed forward by the screw in the pasta press. The area above the pasta extruder screw is where bridging of ingredients can become an issue. The critical factors that can cause bridging are the moisture content of the mix, the level of fill in the mixing chamber, and the properties of the ingredients themselves. Yalla[18] reported that semolina mixed with buckwheat bran flour hydrated at 34% absorption and semolina mixed with flaxseed flour hydrated at 33% absorption were very sticky and bridged during processing. In both cases, bridging was eliminated by reducing the hydration levels. To minimize bridging issues, the mixture should be run as dry as is reasonably possible, and the level of fill in the mixing chamber should be low, but not so low as to starve the screw, which would decrease the output of the pasta press.

Issues with Extrusion of Soy–Semolina Blends

Choosing the correct moisture level for the dough on a pasta press is of critical importance. If the moisture of the mixture is too low, the product will be too viscous, and the press will not have the power necessary to convey the dough and force it through the die. If the moisture is too high, the product will be too thin (low viscosity), and the pasta press will be unable to convey the product forward in the barrel due to slippage at the barrel; this problem can be caused by a minimal variance from the desired moisture content. For example, the rate of extrusion of a semolina–buckwheat bran flour (7:3 ratio) was 30% lower when hydrated to 31% than when hydrated to 30% absorption.[16]

The ideal moisture content is somewhat arbitrary. Variables to control for include good flow characteristics in the mixing chamber, the correct look for the material in the mixing chamber, motor load, die pressure, and maximized output, among others. The authors of this chapter have found that controlling to a desired motor load generally seems to work well. The motor load should be similar to the motor load for semolina or durum flour for the same die and screw speed. Typically, this will also result in a die pressure similar to the die pressure for semolina or durum flour.

The screw on a pasta extruder typically has a small range of effective operating conditions. Because the screw is choke fed, the screw on a pasta extruder is a volumetric device. Lower bulk density of the ingredients results in decreased output of the pasta extruder. Because the output of the pasta extruder has been reduced compared to running pure semolina (due to lower bulk density) and the motor load is controlled to a similar value, heat is created in the dough during extrusion due to the conversion of mechanical energy into heat. This can be demonstrated simply by looking at a calculation for specific mechanical energy (SME). The SME (kJ/kg) transferred to the dough is calculated as the mechanical energy (kJ/sec) required to extrude the pasta divided by the amount of pasta processed (kg/sec). The mechanical energy required to operate the empty press is subtracted from the mechanical energy required to operate the press under load.

Semolina products are typically run near the upper limit of temperatures that still allow quality pasta to be made, typically 45 to 50°C. The onset of protein denaturation can begin at temperatures above 50°C. This means that the temperature of the dough including soy ingredients will be above what would be acceptable for a durum pasta product, which means it is very likely that some of the wheat gluten in the formula will be denatured (as well as possibly initiating gelatinization of the starch). While the water jacket around the barrel will remove some of the additional heat generated, it will not remove all of the additional heat. Limited protein denaturation due to processing can be compensated by increasing the level of vital wheat gluten in the formula.

Some organic soy flours and concentrates have high levels of oils in them. Ingredients that have higher lipid levels would contain correspondingly lower levels of hydrophilic material and may require less moisture to reach proper dough consistency.[16,19] When Manthey et al.[16] worked with buckwheat bran flour having 8% lipid content, they had to reduce the hydration level of the semolina–buckwheat bran (7:3 ratio) 2 percentage points in order to maintain an extrusion rate and a mechanical energy similar to those of semolina.

A pasta press is a single-screw extruder that relies on the barrel wall to provide resistance for the screw to work against. When the level of oil in the formula is too high, the barrel wall becomes lubricated, and the dough rotates in the screw without being conveyed forward. The level of oil that will cause processing problems is dependent on how well the ingredients bind the oil. Formulas can begin demonstrating problems with as little as 2% oil in the total formula or they can be well processed with over 8% oil in the total formula; it all depends on how free the oil is in the system. It appears that the only way to determine when the level of oil becomes an issue in a formula is to attempt to run it. It may be possible to add an ingredient to help bind or emulsify the oil, which would allow for formulas to be processed that otherwise could not be, but information on this approach is limited.

Soy formulas, especially low-carbohydrate formulas, will not process like traditional pasta dough. This may be due to the larger temperature gradients in doughs containing soy ingredients. Also, the viscoelastic properties may simply be too dissimilar from pure durum semolina dough, or additional properties may be affecting the flow. Distribution of flow on the die may be poor and may require corrective measures to make the flow even. Such corrective measures may include blocking some portion of the outlets or redesigning the inserts to force a more even flow.

Typically, soy formulas do not recover from a shut-down very well or very rapidly. This can cause problems when the pasta line is down for intentional or unintentional reasons. When flow stops at the die, soy formulas have a tendency to set up in the die. If flow can be restored, it is typically a slow process. The screw is able to slowly convey the dough forward, frequently running at a motor load that is at or above the production motor load. If fresh product does reach the die, it can be a long time before the flow at all inserts becomes even. In some cases, inserts may not regain flow until the die is cleaned. Finally, it appears that soy doughs can experience changes in volume as the moisture in the dough equilibrates. Under worst-case assumptions, it is possible that the expanding dough could break a die plate or inserts.

Issues with Drying Soy–Wheat Pasta

The addition of soy to the formula appears to benefit some aspects of the drying of pasta. Pasta has a very low moisture diffusivity. The addition of most ingredients (including soy products) tends to improve the drying rate of the pasta. Checking (cracking) of pasta containing soy does not seem to be an issue when the products are dried with a typical pasta drying cycle. Issues with pieces of pasta sticking together almost entirely disappear. The product experiences a very noticeable browning during drying. Freshly extruded pasta may have a golden color, but during drying the color will change to somewhere between a tan and a dark brown. This is likely caused by a Maillard browning reaction. To date, it appears that the issue of the darker color due to drying has not been solved, as no commercial soy pasta samples exhibit the golden color the U.S. market likes to see in a pasta product.

Strategy Two: Structure from Gelatinized Starch

Pasta and noodles that are made without wheat rely on gelatinized starch for structure. Starch gelatinization requires some degree of cooking during processing. These products are often referred to as being precooked. Precooked pasta can be more broadly defined as a pasta product that is cooked prior to reaching the consumer. This definition would include traditional pasta that has been boiled, steamed, blanched, or otherwise cooked. It would also include products where the binding agent is gelatinized starch instead of gluten. For the purposes of this discussion, precooked pasta will include only pastas where gelatinized starch is the binding agent in the pasta. Two basic methods are used to produce pasta and noodles made with gelatinized starch. The first method, the one that is in commercial production in the United States, is to feed a pregelatinized starch into a pasta extruder. The second method, which does not appear to be in current use in the United States, is to use a corotating cooking extruder to cook and form the precooked pasta.

Method One: Pregelatinized Starch

Gelatinizing the starch before extrusion can be done several ways, such as by drum drying, pasting with boiling water, and steaming. During gelatinization, the starch granules swell and amylose leaches from the granule to form a colloidal solution that in turn forms a layer on the surface of granules, resulting in the aggregation of starch granules. Poor starch aggregation would occur with waxy or low amylose starches (0 to <20%).[20,21] In a dough where water is limited, the amylose solution does not dissipate but contributes to the viscosity of the dough. Some starch granules must remain relatively intact.

The process using the pasta extruder is very similar to traditional pasta, with the primary difference being that pregelatinized starch is used as the binding agent instead of gluten. Presumably, the cooked feed could be fed into the mixing chamber either dry (as could be done with a pregelatinized starch) or wet (if a cooking process occurs directly before the pasta extruder). The available information on the process is limited, but some patents have been granted.[22,23] Hauser and Lechthaler[22] patented a process that extruded a non-gelatinized pasta product and gelatinized the starch after the pasta shape was made. The uncooked pasta is able to hold its shape as long as it remains wet; when the pasta is cooked, the gelatinized starch becomes the binding agent that allows the product to maintain its shape through drying and rehydration.

Mestres et al.[24] compared several methods of making pasta products from corn. The corn used in the study was cooked prior to formation of the pasta by several methods: drum drying, extrusion cooking, pasting with boiling water, and steaming the raw corn flour. The study found that the use of monoglycerides in the formula decreased cooking loss by about 50%. This work indicated that the cooking loss for an extrusion cooked corn flour was much higher than for other methods of gelatinizing the starch. Applying a thermal treatment after extrusion (95°C, 95% relative humidity) improved the quality of the pasta.

Non-Wheat Noodles

Making noodles that do not contain wheat protein can be done in one of two ways, although many variations to these two approaches exist, depending on the starch source and ingredients. In the first approach, about 5% of the starchy material is placed in boiling water (1:7) for about 5 minutes, then the remaining ingredients are mixed with the gelatinized portion and warm water is added obtain approximately 40 to 55% moisture. The hydrated mixture is kneaded to form a dough that is generally extruded through a die. The gelatinized portion helps bind the remaining ingredients together during dough formation and extrusion. The extruded noodles are immersed in boiling water for 2 to 3 minutes and then in ice-cold water for 1 to 2 minutes to stop the cooking process, after which they are allowed to drain and the noodle strands are then spread on a rack to dry. Sometimes extruded strands are subjected to a second steaming and allowed to cool to room temperature, then rinsed with water and dried. Alternatively, the extruded strands can be placed in boiling water to fully cook the starch granules, after which they are cooled in cold water, rinsed, and dried. The heat treatment (gelatinization) and subsequent retrogradation determine the stability of starch noodles and their capacity to withstand the boiling water treatment.[20] Rinsing the noodles with cool or cold water enhances the retrogradation of starch. Starch properties affect noodle quality and vary with plant source and within a plant source with regard to genotype and environment. In the second approach, a thick slurry or batter is poured onto a metal sheet which is passed through a steam tunnel; the batter is cooled and cut into noodle strips. The steaming and cooling set the noodle structure. Noodles will fall apart if adequate gelatinization has not occurred.

Method Two: Cooking Extrusion

The second method of production is to use a twin-screw cooking extruder. Much of the work done with twin-screw cooking extruders has been done relatively recently (mostly within the past 30 years), and much of the work has been patented, so a good record of the processes used exists. While quite a number of patents for twin-screw production of a precooked pasta have been granted, the ones cited in this chapter display the basics of the process. Baumann[25] provided an early example of twin-screw production of precooked pasta. Bauman's process was developed to make a pasta similar to traditional pasta but with a shorter cooking time. Semolina was used as the starchy ingredient, but the patent states that other starchy materials can be used. The temperatures in the extruder must reach 90 to 110°C, with a preferred temperature of 90 to 98°C, or they must be sufficiently high to fully gelatinize the starch. The cooked dough is then forced through a die plate to give the dough the pasta shape. The pasta is then dried to a moisture content similar to that of traditional pasta.

Myer and Reid[26] patented a process in which the pasta dough in the extruder reaches temperatures of 235 to 350°F (113 to 177°C). This design uses injections of hot water into the extruder barrel as the source of moisture for the process (it does not use a premixer or preconditioner). Work within the barrel increases the temperature of the dough, and additional heat is provided by heating the extruder barrel to obtain the desired dough temperature. The barrel of the extruder then provides sufficient cooling so the product exiting the dies is cool enough that steam does not flash off and create steam bubbles in the finished pasta. This patent recognizes that many starchy ingredients can be used for the base of the product and indicates the use of glycerol monostearate as an ingredient.

Wenger and Huber[27] patented a process that uses a preconditioner and extruder barrel vent with a vacuum assistance to make a precooked pasta. Both water and steam are injected into the preconditioner to achieve sufficient temperature and moisture content to begin gelatinizing the starch in the product. Additional water and steam are injected into the extruder barrel to increase the moisture and temperature of the extrudate. The vacuum is applied to the barrel at a point after the initial cooking zone, where the screws are not completely filled (limiting issues associated with dough exiting through the vent). This vacuum performs several functions. It allows the product to be rapidly cooled from above 100°C to about 100°C very rapidly by forcing the steam to flash off. This flashing off of steam also reduces the moisture content of the dough which improves the quality of the product at the die and presumably reduces drying costs, as less moisture must be removed from the finished product. It also emulates the traditional pasta process, where a vacuum is applied to the system to minimize the formation of any air bubbles in the pasta. The patent claims that by achieving a high moisture content early in the screw where the shear is the highest for the screw profile used, shear damage to the starch is minimized. Starch that has experienced less shear has greater integrity and results in a final product that has a better mouthfeel.

Pasta and Noodle Quality

Differences between traditional pasta and pasta with soy ingredients can be seen in the data provided in Table 6.1. Traditional pasta has a very yellow appearance (high CIE *b* value), while most pasta containing soy tends to be closer to tan (low *b* value and high *a* value). Other characteristics that are typically measured for pasta include protein content, cooked product weight, cooked firmness, and cooking loss. The protein content of soy-containing samples tends to be much higher than in traditional pasta, as would be expected from the level of protein in the soy ingredients typically used in the formulation of these products. The soy protein tends to have a high affinity for water, so the cooked product weight tends to be high. As a general rule, the addition of non-semolina ingredients into a pasta will disrupt the

TABLE 6.1

Cooking Quality of Commercial Pasta and Noodles that Contain Soy

Sample	Ingredients	CIE Value[a]			Protein Content (%)	Cooked Weight (g)	Cooked Firmness (gcm)	Cooking Loss (%)
		L	a	b				
Spaghetti								
Spaghetti 1	100% semolina	56.49	8.09	44.12	12.5	25.4	7.9	5.1
Spaghetti 2	100% WWD	46.39	10.97	30.12	12.5	25.5	7.7	5.9
Spaghetti 3	SC, SF, VWG, RF, GG	51.24	9.55	30.15	46.4	29.3	9.3	10.3
Rotini								
Rotini 1	100% semolina	49.49	3.91	32.66	12.5	22.1	56.2	4.4
Rotini 2	100% WWD	38.82	8.48	23.32	10.5	20.6	46.6	4.4
Rotini 3	DSF, egg white, RF, VWG, ISP	34.04	10.23	19.39	50.0	26.5	45.5	14.1
Rotini 4	SC, SF, VWG, RF, GG	41.49	8.32	22.43	14.5	24.5	38.9	9.8
Penne Rigate								
Penne 1	100% semolina	40.60	2.44	33.77	12.5	20.7	133.1	3.5
Penne 2	RF, RS, PS, SF	51.01	4.73	25.94	8.9	20.0	25.8	8.6
Penne 3	DSF, egg white, RF, VWG, ISP	38.44	9.74	21.04	50.0	29.0	50.4	17.2

[a] The Commission International de l'Eclairage (CIE) value L is a measure of brightness (1 to 100); *a* is the green–yellow chromaticity coordinate, and *b* is the blue–yellow chromaticity coordinate.

Note: DSF = defatted soy flour; GG = guar gum; ISP = isolated soy protein; PS = potato starch; RF = rice flour; RS = rice starch; SC = soy concentrate; SF = soy flour; VWG = vital wheat gluten; WWD = whole wheat durum.

gluten matrix that forms to retain components that are not molecularly bound to the gluten. This disruption of the gluten matrix allows more of the inert components to leech out during the cooking process. Table 6.1 summarizes the properties of pasta made from durum semolina as well as from other sources. The only trend in Table 6.1 that does not seem to be consistent is the firmness. These data should give a basic indication of the effect that adding soy to pasta has on the properties of the pasta.

Summary

Information available on incorporating soy products into pasta and noodles is limited. The information that is available discusses adding a variety of ingredients into a pasta base (such as whole wheat pasta or the addition of fiber to a pasta product) or discusses the manufacture of a pasta-like product that is made from ingredients other than wheat (such as a precooked rice pasta). These resources can be valuable in helping with the incorporation of soy ingredients if they are viewed from a perspective of pasta as being a carrier for healthful ingredients. Generally, the healthful ingredients will tend to dilute the structurally functional components of the pasta. At some point, this structure must be regained or reduced integrity of the pasta must be accepted. A strategy that works for maintaining the integrity of one product may work well for another; similarities between additives should be explored fully to determine if a solution for a different ingredient can be successfully implemented for a soy ingredient. It must be stressed that, while the product that is being made is, in some sense, pasta, it will never be possible to exactly match all of the characteristics of traditional pasta with a pasta product that is a carrier for more healthful ingredients. Color, flavor, texture, cooking time, and cooking losses, among other characteristics, will change. What is lost in some characteristics is made up for by the improved nutritive quality of the product with pasta as a carrier, which is the reason for adding a product such as soy.

References

1. Marconi, E. and Carcea, M., Pasta from nontraditional raw materials, *Cereal Foods World*, 46, 522, 2001.
2. Grieshop, C.M. and Fahey, G.C., Comparison of quality characteristics of soybeans from Brazil, China, and the United States, *J. Agric. Food Chem.*, 49, 2669, 2001.
3. Grieshop, C.M., Kadzere, C.T., Clapper, G.M., Flickinger, E.A., Bauer, L.L., and Frazier, R.L., Chemical and nutritional characteristics of United States soybeans and soybean meals, *J. Agric. Food Chem.* 51, 7684, 2003.

4. McCue, P. and Shetty, K., Health benefits of soy isoflavonoids and strategies for enhancement: a review, *CRC Crit. Rev. Food Sci. Nutr.*, 44, 361, 2004.

5. Harper, J.M., Macaroni extrusion, in *Extrusion of Foods*, Vol. II, Harper, J.M., Ed., CRC Press, Boca Raton, FL, 1981, p. 19.

6. Borrelli, G.M., Troccoli, A., DiFonzo, N., and Fares, C., Durum wheat lipoxygenase activity and other quality parameters that affect pasta color, *Cereal Chem.*, 76, 335, 1999.

7. Antognelli, C., The manufacture and applications of pasta as a food and as a food ingredient: a review, *J. Food Technol.*, 15, 125, 1980.

8. Pollini, C.M., THT technology in the modern industrial pasta drying process, in *Pasta and Noodle Technology*, Kruger, J.E., Matsuo, R.B., and Dick, J.W., Eds., American Association of Cereal Chemists, St. Paul, MN, 1996, p. 59.

9. Anese, M., Nicoli, M.C., Massini, R., and Lerici, C.R., Effects of drying processing on the Maillard reaction in pasta, *Can. Inst. Food Sci. Technol.*, 32, 193, 1999.

10. Sensidoni, A., Peressini, D., and Pollini, C.M., Study of the Maillard reaction in model systems under conditions related to the industrial process of pasta thermal VHT treatment, *J. Sci. Food Agric.*, 79, 317, 1999.

11. Hou, G. and Kruk, M., Asian noodle technology, in *AIB Technical Bulletin*, Ranhotra, G., Ed., American Institute of Baking, Manhattan, KS, 1998.

12. Hou, G., Oriental noodles, *Adv. Food Nutri. Res.*, 43, 141, 2001.

13. Nagao, S., Processing technology of noodle products in Japan, in *Pasta and Noodle Technology*, Kruger, J.E., Matsuo, R.B., and Dick, J.W., Eds., American Association of Cereal Chemists, St. Paul, MN, 1996, p. 169.

14. Zhang, D. and Moore, W.R., Effect of wheat bran particle size on dough rheological properties, *J. Sci. Food Agric.*, 74, 490, 1997.

15. Manthey, F.A. and Schorno, A.L., Physical and cooking quality of spaghetti made from whole wheat durum, *Cereal Chem.*, 79, 504, 2002.

16. Manthey, F.A., Yalla. S.R., Dick, T.J., and Badaruddin, M., Extrusion properties and cooking quality of spaghetti containing buckwheat bran flour, *Cereal Chem.*, 81, 232, 2004.

17. Sinha, S., Yalla, S., and Manthey, F., Extrusion properties and cooking quality of fresh pasta containing ground flaxseed, in *Proceedings of the 60th Flaxseed Institute of the United States*, Carter, J. Ed., North Dakota State University, Fargo, 2004, p. 24.

18. Yalla, S.R., Effect of Semolina, Nontraditional Ingredients, and Processing Conditions on Spaghetti Quality, M.S. thesis, North Dakota State University, Fargo, 2004, p. 45.

19. Rayas-Duarte, P., Mock, C.M., and Satterlee, L.D., Quality of spaghetti containing buckwheat, amaranth, and lupin flours, *Cereal Chem.*, 73, 381, 1996.

20. Tam, L.M., Corke, H., Tan. W.T., Li, J., and Collado, L.S., Production of bihontype noodles from maize starch differing in amylose content, *Cereal Chem.*, 81, 475, 2004.

21. Bhattacharya, M. and Corke, H., Selection of desirable starch pasting properties in wheat for use in white salted or yellow alkaline noodles, *Cereal Chem.*, 73, 721, 1996.

22. Hsu, J.Y., U.S. Patent Application 327,913, 1981; Rice Pasta Composition, U.S. Patent 4,435,435, 1984.

23. Hauser, T.W. and Lechthaler, J., U.S. Patent Application 654,348, 1991; Process for the Production of Dried Precooked Pastas, U.S. Patent 5,122,378, 1992.

24. Mestres, C., Colonna, P., Alexandre, M.C., and Matencio, F., Composition of various processes for making maize pasta, *J. Cereal Sci.*, 17, 277, 1993.
25. Baumann, G., U.S. Patent Application 609,075, 1975; Process for Preparing Precooked Pasta Products, U.S. Patent 4,044,165, 1977.
26. Myer, W.J. and Lamppa, R.G., U.S. Patent Application 566589, 1983; Multiple Screw Pasta Manufacturing Process, U.S. Patent 4,540,592, 1985.
27. Wenger, M.L. and Huber, G.R., U.S. Patent Application 111,196, 1987; Low Temperature Extrusion Process for Quick Cooking Pasta Products, U.S. Patent 4,763,569, 1988.

7

Soy Base Extract: Soymilk and Dairy Alternatives

Ignace Debruyne

CONTENTS

Introduction

Soybeans can easily be transformed into many vegetarian dairy alternatives based on whole soybean powders or on soy base extracts. Alternatively, the products can be made using a reconstitution formulation. The focus of this chapter is on the optimization of soy base extraction technology. Using soy base as a raw material creates opportunities to develop a vast range of value-added products, such as drinks, smoothies, tofu, yogurt, desserts, and ice cream. Using the whole soybean as the starting material provides great flexibility in downstream processing where the number and amount of natural soy components must be maintained at levels required for specific products. Soymilk, as typically defined, is a product made directly from soybeans, although modern soymilk standards also include drinks based on soy protein isolates or soy powders combined with soybean oil.[5,15] It has been made for thousands of years in China and neighboring countries. The traditional method of soymilk production essentially contains four consecutive steps:

- The soybeans are soaked in water, usually overnight and at room temperature, to allow the soybeans to swell.
- The soaked soybeans are ground into slush.
- The ground soybean slush is cooked.
- The soy base extract (soymilk or tonyu) is separated from the soy fiber fraction (okara).

Traditional Chinese-style soymilk has a distinct soy taste, with lots of beany and other green flavor notes, which are highly appreciated by Chinese and other Asian consumers but not preferred attributes for the majority of consumers in the rest of the world.

Dry and Semidry Processing

Soy base extract and soymilk can be based on the whole soybean, but also on full-fat soy flakes or full-fat soy flour. Some processors prefer using preprocessed soy, thus reducing tedious in-house soybean cleaning, dehulling, and preparation. One example of such a product is the soy flakes produced by MicroSoy Corp. (Jefferson, IA).[9] Flakes intended for soymilk or tofu production are based on clean, dry, whole soybeans, preferably low in fiber, full fat, and enzyme active with a thickness of around 200 µm. These characteristics of the flakes allow rapid rehydration compared to whole soybeans (5 minutes vs. many hours). The production method utilizes mechanical processing, including high-speed color sorting, cleaning and dehulling, and flaking. Another example is the ultrafine whole-bean soy flour obtained using the Buhler Micromill process.[3] The soybeans are first dehulled using Buhler's hot dehulling process, and the cotyledon particles are then micromilled to a free-flowing powder by means of a six-roller mill, equipped

FIGURE 7.1
Buhler Ultramill. (Courtesy of Buhler Group, Uzwil, Switzerland.)

with proper cooling to avoid local heating and denaturation of the flour particles. The hot dehulling is said to reduce bitterness formation. The mill grinds the material by means of three sets of rolls, with one roll in each pair rotating faster than the other (Figure 7.1). The great forces created in the narrow roll gaps shear the tough kernels apart, rupturing the plant cells and releasing their contents. In a first step, microflakes with a thickness of 10 to 50 μm are obtained; on the next set of rolls, the plant cells are further broken down, liberating the protein bodies and fat globules and making them easily available for extraction.

Other processes that are on the market include the use high-speed pin mills or jet mills in combination with air classifiers or grinding under nitrogen. Ultramilling gives soy flour with a typical particle size of less than 10 μm (1000 mesh sieve), or less than 30 μm (400 mesh sieve). Such flour is readily dispersible in water and requires little or no extra homogenization and no or limited fiber separation.[3] The risk for off-flavor formation, however, remains high, as full-fat flour is highly sensitive to oil oxidation.

More recently, Archer Daniels Midland, for example, launched instant soymilk powders based on heat-treated and spray-dried whole soybean extracts. The process promises proper inactivation of the enzymes that could induce beany flavor in the end product. Dry soymilk powders, though, remain very sensitive to oxidation and will always require proper handling and storage conditions to reduce the risk of off-flavor development.

Soy Base Extraction Process

The objective of preparing soymilk is to overcome deficiencies of the traditional product. A good soy base extraction process should, therefore:

- Extract the soluble protein as much as possible.
- Achieve the desired flavor, taste, and texture.
- Inactivate enzymes in time to avoid side reactions.
- Stabilize the product for longer shelf-life.
- Achieve an excellent taste, the number one consideration.
- Guarantee the nutrition value.
- Be affordable and operate economically.

Soy base extraction technology has evolved considerably over the past 30 years.[1,6,7,11,13] New developments have greatly improved the taste of the basic product. A bland soy base leads to greater acceptance of soy products, promotes production of the next generation of dairy alternatives, and allows reproduction of the same taste wherever and whenever the product is made.

Flavor Characterization of Soy Dairy Products

The definition of a good product or process may differ with the point of view. For production unit management, the process should increase company profits and improve the bottom line, and the finance department would be primarily interested in least-cost formulas, a consistent flow of supplies from vendors, and increased sales activities. But, the ultimate goal of any process must be to satisfy the consumer's preference for a light-colored, low-cost product that has a bland taste (beany taste in Asia), no grittiness, and little aftertaste. Ideally, the soy would not be genetically modified (GM), and organic would be preferred. An acceptable process or product would have the desired:

- Sensory characteristics
- Nutritional properties
- Physicochemical properties
- Microbiological stability
- Functional characteristics

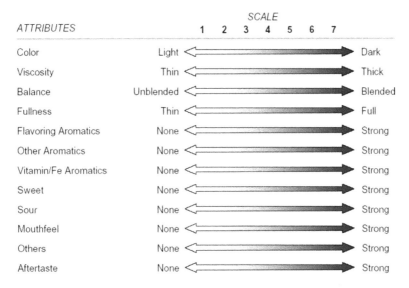

FIGURE 7.2
Critical sensory attributes.

Critical sensory attributes for soy-based drinks were identified in the Soyatech and Arthur D. Little study of 2001.[12] The study also gives an excellent review of how to organize a sensory evaluation (Figure 7.2). Setting and evaluating these sensory attributes is by definition a subjective business; however, some major variables can be defined (Figure 7.3):

- Color (white to yellowish), which is determined by soybean quality characteristics, as well as emulsion properties, which are soybean quality and process dependent
- The presence and coarseness of particles, which are primarily determined by the homogenization and separation efficiency
- Bitterness, which is determined by several factors, including the presence of isoflavones (aglucones), saponins, phytic acid, and occasionally the presence of bitter peptides, byproducts of protein hydrolysis
- Sweetness, which can be obtained by using sweet soybeans or easily be adjusted by adding sweeteners or sugars
- Beanyness and related taste characteristics, which are the product of lipoxygenase action on polyunsaturated fatty acids (PUFAs) and can be prevented by proper enzyme denaturation and further reduced by avoiding oxidation due to processing (similarly, rancidity development should be controlled properly)

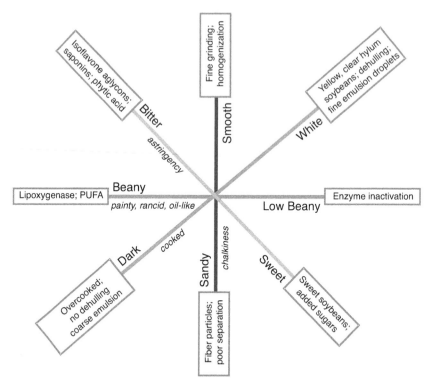

FIGURE 7.3
Sensory characteristics rose.

Off-Flavor Development and How To Control It

Chemical analyses of off-tastes in soybeans have revealed several unwanted components that are linked to specific flavors:

- Green or beany taste — grassy aldehydes, beany furans
- Fishy taste — volatile amines
- Bitter taste — oxidation products, fatty acid dimers, trihydroxy fatty acids
- Astringent taste — phenolics
- Roasted taste — browning products
- Oxidized taste — grassy/beany alcohols
- Cereal taste — furfurals

Most of these flavors are linked to the oxidative breakdown of oil and protein components, although bitterness and cooked or roasted notes may have a different origin. The main culprit behind oxidation reactions is the lipoxygenase (LOX) enzyme, several types of which have been identified in regular soybean varieties. LOX is a catalytic protein that requires a convergence of all substrates under specific conditions to create hydroperoxides. To do so, LOX requires both PUFAs and oxygen. PUFAs are abundantly available in soybean oil. Preventing the actions of LOX therefore focuses on avoiding contact with air, especially in aqueous conditions:

$$H_2O + O_2 + PUFA \xrightarrow{\text{lipoxygenase (LOX)}} \text{fatty acid hydroperoxides}$$

These hydroperoxides will be turned into volatile off-flavors in a second reaction that is catalyzed by the enzyme hydroperoxide lyase:

$$\text{Fatty acid hydroperoxide} \xrightarrow{\text{hydroperoxide lyase}} \text{short – chain aldehydes}$$

Other reactions result in the formation of ketones, ketols and other alcohols, furans, and hydroxy acids.

Avoiding oxidative breakdown can be accomplished in two ways: (1) reducing or even blocking access to oxygen or air, or (2) inhibiting or simply denaturing the key player in the process, the LOX enzyme. Fortunately, LOX is one of the least stable proteins in the soybean, so LOX activity can be easily denatured by proper heat treatment. Several methods for LOX denaturation or inhibition have been proposed:

- Heating — Protein denaturation is a function of time, temperature, and moisture level; LOX denaturation, especially at a higher moisture content, is rather simple (a few minutes above 55 to 60°C at a high water content is more than sufficient).
- pH control — Enzymes are highly pH sensitive, but the acceptable pH range in products might be limiting.
- Substrate removal — Reduce access to oxygen (or air).
- Remove reaction products (thus eliminating the problem).
- Enzymatic approach — The UniCell® process developed by Cellfoods in Japan utilizes enzymatic dehydrogenase breakdown of hexanal and hexenal.[14]
- Fermentation — Utilize microbial breakdown and microorganisms having sufficient dehydrogenase activity.

It is better to avoid taste deterioration than to correct it when damage is done. In practice, a combination approach is highly recommended: (1) reduce access to oxygen during the entire process, and (2) denature the LOX activity as early as possible in the processing procedure.

Bitter taste and astringency are directly linked to the presence of isoflavones, specific (bitter) peptides, and perhaps the presence of phytic acid and a low level of free fatty acids. The isoflavone aglucones are particularly known for their astringency. Isoflavones are present naturally as esterified glucosides. Isoflavones are also concentrated in the soy hypocotyl (soy germ), which contains around 50% of the total isoflavone content of the beans (mainly the daidzein form). Removing the germ or simply selecting low-isoflavone varieties might help. Avoiding isoflavone glucoside hydrolysis is the next approach to controlling bitterness. Other options are process control (reducing isoflavone hydrolysis to the aglucon form; reducing protein hydrolysis), debittering (isoflavone or germ removal; soybean leaching), and masking the bitter taste.

Heating the soymilk is necessary at some point in the process. The main objectives are improving the nutritional value of the soy protein by reducing antinutritional factors such as antitrypsin and improving the protein digestibility corrected amino acid score (PDCAAS). For good digestibility, the protein must be properly denatured; however, for taste reasons, heating should be done as early as possible in the process. Extra cooking for nutritional reasons and for product stabilization can be done at the end of processing.

Heat denaturation of protein that is too intense can have unwanted side effects. Protein fiber interactions might reduce the yield, too much heating might compromise taste, and, of course, heating can be costly. Native soluble proteins easily dissolve in water and guarantee a better taste, better nutrition, better yield, and lower formulation costs with fewer calorie input.

Off-flavors should be prevented as much as possible, if not completely; however, correction is possible during the process by adding a flashing step (see below) or afterwards by utilizing targeted flavoring. Flavoring specialists will primarily use three different types of flavors:

- Potentiators, which increase the perceived intensity of the flavor
- Enhancers, which increase the pleasantness of the flavor
- Modifiers, which enhance, suppress, or change the flavor of the food

Soybean Selection

The best soybeans for modern dairy alternative products are those with fewer flavors and subtle creaminess, thereby eliminating the beany taste and improving the overall texture. Many varieties of soybeans are available for feed and food. Different growing regions and climates require specific agronomic traits for optimal yield. The crushing industry mainly uses commodity soybeans. The food ingredients and soyfoods industry will mainly work with identity-preserved (IP) specialty soybeans. IP soybeans can be variety or genetic trait specific (e.g., Vinton 81, Beeson, Sunrise, Laura), can have desired processing characteristics (unique protein-to-oil ratio, protein type, oil type, lipoxygenase level), can have particular physical attributes (large or small; yellow, black, or clear hilum), or can come from a specific growing region.

Additional production process characteristics can be required, such as non-genetically modified IP (non-GM IP), GM-IP (varieties developed using genetic modification for specific end uses, such as improved food quality or protein or oil characteristics), or organic IP (non-GM soybeans grown according to accepted standards for organic cultivation and processing).

Proper moisture control and selecting fresh soybeans harvested at the ideal maturity will help ensure good product quality. Soyfood soybeans are always cleaned of all impurities, foreign material, and splits or broken beans (so-called grade 0 quality). Storage and handling conditions are equally important: at the farm, the cleaning and grading unit, during shipping, and at the soymilk processor site. Proper control of temperature, humidity, air circulation, and product load is essential.

Soybean Preparation

Soybeans used for food should comply with basic quality specifications. If not yet done, the soybeans should first be cleaned to the highest possible standard. A proper soybean cleaning and preparation unit will combine a destoner with a gravity table and screening, sifting, and polishing units to obtain soybeans of equal size with no impurities, foreign material, or splits and broken soybeans left.[6,7] Optionally, color sorting can be used to select soybeans of one specific grade of color, removing black, purple, green, and mottled beans. Soybeans with light and consistent color will yield a whiter soy base. For similar reasons, some producers prefer using clear-hilum soybeans, as the presence of remaining black-hilum particles can yield unwanted effects in soybean curd (tofu) made from such soymilk.

Soybean Dehulling

Preparation optionally includes a dehulling step. If applied, the dehulling operation should be part of an integrated effort to reduce off-flavors, eliminate bitter tastes, and improve the emulsion stability of the final product. Large-scale dehulling in a crushing plant entails soybean preconditioning followed by utilization of breaker rolls and hull separation, a technique that yields only partially dehulled soybeans and lots of breakage, increasing contact with air and the risk of oxidation. For soy base extraction, dehulling must be done more carefully in a small-scale cold or hot dehulling unit.

Hulls can absorb large volumes of water, leading to clogging of the processing equipment and affecting the overall production process; eliminating hulls also reduces the amount of okara byproduct. In small-scale processing, though, hulls can help in the filtration step. Cold dehulling simply removes the hulls and yields clean soybean halves or cotyledons. Properly executed hot dehulling eliminates or at least reduces the soybean LOX activity, which should result in a higher quality soy base and better tasting tofu. Splits and damaged beans are easier to remove. Dehulling also facilitates the removal of isoflavone-rich soy germs, a step that positively affects taste by reducing the potential for bitterness and astringency.[6,7]

An additional bonus of hot dehulling results from changes in the soybean lecithin composition that improves the emulsion stability of the soymilk.[2] Soybean lecithin is the natural emulsifier of soymilk; no extra emulsifiers should be used. Native soybean lecithin, however, is better suited for water-in-oil emulsions, but it tends to destabilize soymilk during long-term storage. Proper treatment during hot dehulling can trigger the release of phospholipase D, thus yielding more phosphatidic acid, which will improve the oil-in-water emulsion stability of the final product.

Typically, hot dehulling is a combination of fluid bed heat treatment (e.g., using the Buhler OTW-Z unit), an impact mill, a working unit with a paddle-mixing system, and separation steps (e.g., using Kice air separators).[3] At the end of the process, sieves remove unwanted splits. Proper vapor or water addition and moisture equilibration before fluid bed treatment must be combined with a targeted heating gradient to trigger the *in situ* enzymatic lecithin breakdown.[2]

Soybean Soaking and Blanching

Making soymilk requires water. For industrial operations, the water should be soft and clean. If city water is used or the water is pumped from a well, proper pretreatment is required: softening, perhaps demineralization if the water is too high in salts, and of course filtration to remove impurities. A high salt content can considerably affect taste and yield. Calcium ions especially will bind to soy protein and form tofu-like curd particles that will destroy the general appeal of the final product by yielding a deposit in the packed product. A common procedure for making soymilk begins with washing, then soaking the soybeans in water for 3 hours or more, followed by separation of the soybeans from the water. Next, the beans and newly added water are ground into a slurry. This slurry is heated to boiling for 15 to 20 minutes to sterilize the product and improve nutritional value and flavor. These steps can be modified in order to better control off-flavor formation and facilitate grinding and processing.

Skipping the soaking step is the ultimate simplification; however, it will quickly cause major problems in the grinding equipment, as soybeans are notoriously abrasive. Wear and tear of the grinder will be obvious, even when special metal alloys are used. Soaking the beans in cold water induces only limited or no LOX reactions, especially when done with minimal air contact and movement. Coldwater soaking requires much longer time, though, than soaking in lukewarm or in hot water. Soaking in hot water has the additional advantage that it will eliminate all negative enzyme reactions very rapidly. This is why many processes include a hot blanching step. Typically, 15 to 20 minutes of blanching in three to five volumes of water at

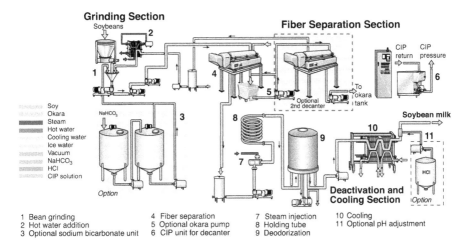

Grinding Section

Fiber Separation Section

Soybeans

CIP return | CIP pressure

Soy
Okara
Steam
Hot water
Cooling water
Ice water
Vacuum
NaHCO₃
HCl
CIP solution

NaHCO₃

To okara tank
Optional 2nd decanter

Soybean milk

Deactivation and Cooling Section
Option

Option

1 Bean grinding
2 Hot water addition
3 Optional sodium bicarbonate unit

4 Fiber separation
5 Optional okara pump
6 CIP unit for decanter

7 Steam injection
8 Holding tube
9 Deodorization

10 Cooling
11 Optional pH adjustment

FIGURE 7.4
A full soy base extraction process. (Courtesy of Tetra Alwin™, Singapore.)

85 to 90°C is used; the equilibrium temperature inside the soybeans or cotyledons should reach 80°C or more. Treatment that is too long or too hot could reduce protein solubility and yield. This effect is limited, though, because at the end of the process all protein will be present as insoluble colloid gel particles anyway.

In the Soya Technology Systems and University of Illinois process, adding sodium bicarbonate to the process water in the blanching or grinding step is recommended.[13] Blanching of the soybeans in a sodium bicarbonate solution at high temperature can inactivate the LOX enzyme; it is also supposed to improve protein solubility by inducing a slightly more alkaline pH. This may be true, but bicarbonate addition is not necessary to make high-quality soymilk (see optional unit 3 in Figure 7.4). It also complicates the process by adding chemicals; depending on the water quality, extra neutralization with acids might be required at the end of the process.

Soaking and blanching are like washing, as these steps remove dirt from the soybeans; however, some loss of soluble components occurs as well. Salts, flatulence-causing oligosaccharides, and other water-soluble components such as isoflavones will partially leach from the soybeans or cotyledons. This has direct taste effects in that less salt and sugar results in a flatter, less complete taste profile, whereas fewer isoflavones reduce the risk of bitterness and astringency development. Continuous blanching is carried out in a countercurrent blanching unit, such as used in vegetables processing. The screens and transport system have to be optimized for soybean or soy cotyledon size. Hot blanching is preferably done with dehulled cotyledons rather than with whole soybeans. Hulls tend to loosen during the blanching and can clog the process unit. Novel enclosed blanching equipment units are being developed that could eliminate the risk of equipment clogging.

Soybean Wet Grinding

The major objectives of the grinding step are opening the plant cells to release the lipids and protein bodies into the water, thus creating a protein-rich, oil-in-water emulsion, and creating a slurry that can be easily separated in a fiber-rich okara with as little water as possible and a soy base with as high a protein content as possible. This objectives can be achieved by maintaining a proper water/soybean ratio and using appropriate grinding equipment. Grinding should not leave any visible or detectable soybean particles. Particles indicate clumps of unopened cellular material, hence a loss in yield and valuable product. After soaking or blanching, the water can be removed and replaced with fresh process water. If heating has not yet been done, hot water is used or steam is injected to increase the temperature of the soybean slurry as quickly as possible.

Grinding dry soybeans in cold water is not a good idea, as it will only damage the equipment and reduce the product yield. In small-scale tabletop equipment such as the ProSoya VS40, steam is injected in the batch reactor containing the presoaked soybeans and the water, while the machine grinds continuously. In a continuous process, the presoaked soybeans in the fresh process water can be subjected to direct steam injection, which induces a temperature rise to 100°C or more in just a few seconds. The ProSoya process, for example, is based on the principle of airless grinding: Presoaked soybeans or cotyledons are mixed with the process water and then pass through a closed high-speed screen grinder. In larger-scale production units, such equipment is only the first step of the grinding process.[10]

After hot blanching, the cotyledons are preferably mixed with hot water. Colloid milling (e.g., Fryma Maschinen) will follow screen grinding to optimize particle size reduction and cell opening (see units 1 and 2 in Figure 7.4). Too strong cooking before the grinding can reduce the protein solubility and extraction yield; however, in most cases, inefficient grinding is the cause of poor yield. This is why in some cases high-pressure homogenization is the final step in slurry homogenization. Homogenizing at 150 kg/cm^2 will guarantee complete cell opening; higher pressure can be used but is costly and will yield too great a reduction in the size of the cell wall fiber material, making the fiber separation difficult if not impossible.

Flash Deodorization

Some soymilk process technology suppliers (e.g., Tetra Alwin™, APV, and ProSoya) offer an optional flash deodorization treatment (see units 7, 8, and 9 in Figure 7.4). Flash deodorization will help reduce any off-flavors already present in the intermediate product; however, proper processing can eliminate such expensive treatment. Flavor correction can also be done later in the process and may be part of the flashing used in direct ultra-high-temperature (UHT) treatment, which is the last step before packaging the product. When using direct UHT, a flash treatment earlier in the process would be redundant. Flash treatment requires a closed-loop system adapted for high-pressure steam injection. Food-quality steam is injected directly into the slurry coming from the grinding equipment. The temperature rises to 125°C or more in just a few seconds. The product is then pumped through a holding tube, where the high temperature can be maintained for some time, typically 5 to 15 seconds, before it is sprayed into a vacuum kettle, where the overpressure is lost. Off-flavors are removed together with excess steam and water, and the product cools rapidly to around 80°C.

Flash deodorization can be applied to the entire slurry (e.g., ProSoya™ process) or just the soy base itself (e.g., APV or Tetra Alwin™ process). Flashing more viscous slurry that still includes the okara complicates things and increases the risk for clogging the flash nozzle; however, it has an economic advantage in that it avoids reheating and yields food-grade cooked okara. If this is the first heating step in the production process, then the deodorization step can serve as the enzyme-inactivation step for LOX, trypsin inhibitor, and other enzymes such as urease, proteases, and lipases (which may cause flavor deterioration in long-shelf-life products), with no loss of insoluble protein, as the heat treatment is after the grinding and separation.

Okara and Soy Base Separation

In a tabletop batch process, after cooling below 100°C the slurry is filtered on cheesecloth that separates the soy extract (tonyu) from the insoluble fiber (okara). This is the simplest separation technique possible and can be quite inefficient due to the lack of proper pressing of the cake. In small-scale industrial lines, the slurry is filtered on one or more scraped filters in batch or semicontinuous mode (e.g., Embrasoy, Perfecta Curitiba, and Agrolactor systems). The quality of the okara is an indicator of grinding efficiency; the collected okara should not contain any soybean particles that can be detected visibly or by touch.

A good okara also should contain as little water as possible (actually, the water represents lost soy base). Okara is extremely rich in fiber that binds large volumes of water. A dry matter content of 22% is a very good start. The volume of sediment (20% or more) to be removed is too high for the average industrial disc stack centrifuge. In continuous industrial production units, the separation must therefore be done with a decanter (see unit 4 in Figure 7.4). Soy base belongs to the "soft" product category and requires a special decanter design for adequate processing.

A properly designed decanter (Figure 7.5) processes a high capacity of soymilk while producing a relatively dry okara cake. Okara is removed using a special okara discharge pump. Many manufacturers offer decanters optimized for okara separation (e.g., Westfalia, Tetra Alwin™, Flottweg, Andritz Guinard).

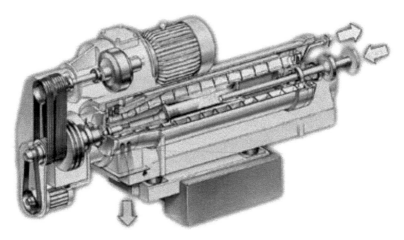

FIGURE 7.5
Soy base decanter. (Courtesy of GEA Westfalia Separator, Northvale, NJ.)

Okara Washing

The volume of okara to be separated is significant (1.4 to 1.8 kg okara per kg of soybeans). Okara is high in moisture content (75% or more), and, as expected, a considerable loss of soy base in the okara fraction occurs in a one-step separation process. In a large-scale soymilk production, this is not acceptable. A solution is countercurrent okara washing. Okara washing is difficult to realize in a continuous countercurrent reactor–separator design. Okara is too soft and has a limited density difference with the continuous water phase. Washing is therefore realized in separate units integrated in a continuous production system. In industrial operations, a single washing step will produce a yield increase of more than 10% extra soy base extract. A single washing is usually considered sufficient.

In a first decanter, the okara separation is fine-tuned to obtain the maximum amount of clean soy base with no okara particles at all. The outgoing okara is immediately dynamically mixed in-line with fresh process water and pumped into the second decanter or dedicated separator (see unit 5 in Figure 7.4), which is optimized for the production of very dry okara. The outgoing washing water (the presence of some okara particles does no harm) is used together with extra fresh process water for blending with the cotyledons arriving from the soaking or blanching step, just before the wet grinding (unit 2 in Figure 7.4).

This design not only guarantees a considerable overall yield gain (10 to 12% more soymilk at the same protein concentration) but also permits working at much lower process water levels. The result is a soy base extract with a considerably higher dry matter content (Table 7.1), providing more flexibility in downstream processing into other products such as tofu and soy yogurt.

TABLE 7.1

Solids and Proteins on Wet Basis (Averaged Data)

	One-Decanter	Two-Decanter
Total solids	8.5	10.5
Protein	4.0	5.0
Fat	2.5	3.1
Soluble sugars	1.0	1.3
Other	1.0	1.1

Cooling, Storage, and Production of Dairy Alternatives

After separation and optional deodorization, the soy base extract is cooled on a plate-heat exchanger before intermediate storage (see unit 10 in Figure 7.4). This soy base should have around 4% protein when produced on a single decanter and up to 5% when produced via two-decanter countercurrent okara washing (Table 7.1).

Okara Byproduct Valorization

Okara is a sticky high-moisture product. Okara contains primarily soy cell wall fiber and, if not dehulled, soy hull material, as well as soy base or okara washing water. As such, it is a very perishable product due to its high soluble carbohydrate and protein levels. It will quickly acidify due to wild or directed lactic fermentation and stabilize as a pumpable feed ingredient for swine rations, for example. For this application, okara can be mixed with the washing or blanching water exiting from the soaking or blanching step; however, this is a very low-value application of a product with great potential as a functional food ingredient, so many alternative application routes for okara have been screened with varying success.

Direct application is possible in bread and bakery products. This can be achieved easily in small-scale soymilk production, but the logistics for short-term application are quite a nightmare in large-scale production units. Drying is costly because more than 70% of the water must be removed from fiber that specializes in water absorption. Drying yields a white or yellowish powder that still has considerable soy fiber functionality (i.e., water-absorbing properties). Powdered okara can be a functional carrier in feed formulas, but it is most of all an excellent and healthy food ingredient that increases the bound water content in a multitude of food products. To improve drying efficiency, okara can first be liquefied chemically — for example, acidic hydrolysis under pressure cooking conditions yields the soluble soy protein stabilizer (SSPS) marketed by Fuji Oil[8] — or enzymatically using α-galactosidase, which yields a yogurt lookalike. Hydrolyzed okara or fractions thereof have been utilized as food additives or stabilizers.

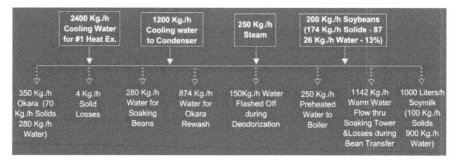

FIGURE 7.6
Mass balance for the ProSoya VS 1000 soymilk plant (1000 L/hr at 5% protein level). (Courtesy of ProSoya, Ottawa, Canada.)

Yield Calculation

With 1000 kg of soybeans having around 37% protein, full dehulling, and a single decanter system, about 6800 L of soymilk at 3.4% protein can be produced (8250 L at 2.8% protein), together with around 1100 kg of okara (at 20% dry material). With a two-stage decanter process, the soymilk yield increases about 10%. The soy solids yield (Y_S) and soy protein yield (Y_P) can be calculated according to the following formulas:

$$\text{Soy solids yield:} \quad Y_S = \frac{S \times N \times (1 - L / 100)}{B \times (1 - M / 100)}$$

$$\text{Soy protein yield:} \quad Y_P = Y_S \times (Q / P)$$

where S = soymilk produced (kg), N = solids in soymilk (percent measured with moisture balance), L = dehulling and cleaning loss (%), B = soybeans (kg), M = moisture (%), Q = protein in soymilk (percent on a moisture-free basis), and P = protein in soybeans (percent on a moisture-free basis). (See Figure 7.6.) Because soymilk extraction aims at protein extraction, the protein yield should always be superior to the solids yield, as illustrated in Table 7.2. Removing blanching water reduces the solids yield while only slightly eroding the protein yield. Introducing countercurrent washing (no blanching) improves the protein yield.

Soy Base Extraction Equipment Suppliers

Equipment for soymilk production should be tailored to the customers' priorities. Small-scale users focusing on the over-the-counter (OTC) market of soymilk and food products based on soy base, okara, and tofu will be served very well with a table-top production unit such as the ProSoya VS40

TABLE 7.2

Soy Solids and Protein Yield Calculations in Soymilk Production

Yield	No Blanching		Blanching		Two-Stage Decanter	
Soy Solids Yield (Y_S)						
Soybeans (*B*)	100 kg		100 kg		100 kg	
Moisture (*M*)	12.0%		12.0%		12.0%	
Soymilk produced (*S*)	680 kg		670 kg		750 kg	
Solids in soymilk (*N*)	7.0%		6.5%		6.5%	
Soy Protein Yield (Y_P)						
Protein in soybeans	32.6%		32.6%		32.6%	
Protein in soybeans (on DM) (*P*)	37.0%		37.0%		37.0%	
Protein in soymilk	3.4%		3.4%		3.4%	
Protein in soymilk (on DM) (*Q*)	48.6%		52.3%		52.3%	
Drilling and Cleaning Loss (L)	Y_S	Y_P	Y_S	Y_P	Y_S	Y_P
0%	54.1	70.9	49.5	69.9	55.4	78.2
7%	50.3	66.0	46.0	65.0	51.5	72.7
11%	48.1	63.1	44.0	62.2	49.3	69.6
15%	46.0	60.3	42.1	59.4	47.1	66.5
18%	44.4	58.2	40.6	57.3	45.4	64.1
21%	42.7	56.0	39.1	55.2	43.8	61.8
24%	41.1	53.9	37.6	53.1	42.1	59.4

or other version (e.g., SoyCow, Soya Goat). Such an approach provides an opportunity for establishing microbusinesses in developing countries or can be used in school cafeterias, for example. When volumes of 200 L per hour or greater are desired, batch or continuous production units are preferred. ProSoya (Figure 7.7), SSP India, Assoy, Embrasoy, Perfecta Curritiba, and Agrolactor of Nutritech and Bar N.A. are just a few of the many players in this market niche. In this capacity range, it is important to monitor the efficiency of the grinding step and the separator. Grinding should not leave any detectable particles; the separation should be easy and yield the maximum dry matter possible with a soy base free of remaining particles. Several units currently in operation have failed on one or both these requirements. This level of production requires product formulation, stabilization (mostly pasteurization), and packaging.

Large-scale production (above 1000 L per hour) requires long production runs (up to several days), excellent quality, excellent yield, and, of course, the lowest production costs possible (see Figure 7.7). Setting up production can be realized in various ways, such as doing it yourself or buying a turnkey plant. Reinventing soy base extraction may occur once again. This process has contributed to the success of such soymilk producers as Silk and ProSoya in North America and Alpro, Sojasun Technologies, and Natumi in Europe. These companies at one time also offered this technology to the market;

FIGURE 7.7
ProSoya VS200C. (Courtesy of ProSoya, Ottawa, Canada.)

however, developing such technology is not a core business for soymilk producers. Many equipment vendors offer complete production units or important subunits of the production process optimized to a particular customer's needs. Some major producers of turnkey plants include ProSoya of Canada and its licensee SSP India, APV, Tetra Pak (Tetra Alwin™ process), and Takai of Japan. Many companies offer grinding, homogenization, and separation equipment (e.g., APV, Westfalia, Fryma, Flottweg, NIRO, Rossi & Catelli). Cleaning-in-place (see unit 6 in Figure 7.4), formulation, sterilization, and packaging equipment are the same as elsewhere in the food industry.

Stabilization and Packaging

Pasteurization (Figure 7.8) can be sufficient for medium-shelf-life products. By definition, it is combined with cold storage for a period of up to 8

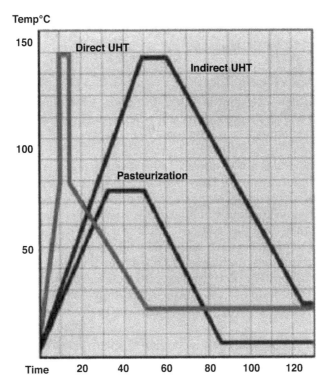

FIGURE 7.8
Temperature profile for stabilization processes.

(sometimes 12) weeks. In recent years, a major soy beverage marketing breakthrough in the United States and United Kingdom was realized with the introduction of pasteurized soy drinks, which can be positioned next to dairy milk, which is found in cold shelf storage areas in supermarkets. Pasteurization is also the first choice for lower production capacity units in developing markets, where the product can be packaged and sold in plastic sachet or low-price bottles or boxes.

Other markets prefer the sterilized or UHT- or high-temperature, short-time (HTST)-treated product packaged in fancy bottles or bricks for the high-end retail market. UHT treatment of soymilk products typically is done using direct UHT with steam injection followed by a short holding time of around 5 seconds at about 145°C, followed by flashing to remove off-odors and excess steam. Some producers prefer indirect UHT, but the risk of cooked flavor development in combination with the potential to retain off-flavors is limiting (for temperature–time profiles, see Figure 7.8). More recently, infusion UHT treatment has been made available as a valid alternative. Details on the UHT process and various packaging technologies are beyond the scope of this chapter.

Reconstituted Soymilk

For the formulation of dairy alternatives, regular food industry formulation equipment should do. Reconstitution uses soy protein ingredients such as soy protein isolates, concentrated soy protein, or even defatted soy flour in combination with a range of other ingredients, such as oils or fats, various types of carbohydrates, salts, one or more emulsifiers and stabilizers, coloring agents, and flavors. The food products can easily be formulated, so soybean extraction is not required and byproducts are avoided. Full-fat and defatted soy flours are major ingredients in low-cost dairy alternatives. These replacements are used in beverages for human consumption in several developing countries. With the development of soy protein isolates, higher quality soy-based infant formulas and soy drinks for the general market became possible. These products have improved color, flavor, and odor and do not contain the flatulence-producing carbohydrates found in soy flour or soy base. Such soy protein products can also be used to increase the protein content of a multitude of food products outside the range of dairy alternatives. Instant soymilk powders, a special niche in this market, are based on:

- Isolated soy protein, combined with vegetable oil or fat, emulsifiers, and other ingredients
- Spray-dried soymilk or soy base, combined with other ingredients
- Spray-dried, enzyme-treated soy base (for which fiber hydrolysis leaves the okara in the product), combined with other ingredients

Soy protein isolates and spray-dried soymilk powder compete against each other in this market. Major players in the soy protein isolate market are the Solae Company, Archer Daniels Midland, Cargill, and Solbar Plant Extracts.

Soy Dairy Alternatives

Following the commercial success of soymilk, dairy alternatives based on soymilk or mixtures of soy and cow's milk have been developed. For individuals seeking alternatives to cow's milk, cheese, yogurt, and ice cream, a growing number of such foods comes from soy. These products have been well accepted by vegetarians and the cholesterol-conscious population.[4,6,7] Soy base is the perfect starting material for such dairy alternatives, and soy base with a higher protein content adds more flexibility for subsequent processing. Soymilk is both a component and an end product. It can be formulated into a large number of plain or flavored beverages of varying protein content.

Plain soy base works best for sauces, gravies, soups and other savory dishes. The lighter colored soy base retains the light color desired for many desserts and sauces. Culinary experts usually prefer cooking with whole soy base rather than soy protein ingredients to obtain a firmer consistency in cooked dishes such as puddings or custards. Many companies produce soy and milk protein blends that are sold as ingredients to food manufacturers. These blends are used as complete or partial replacements for nonfat dry milk in baked goods, sauces, meat products, and various fabricated foods.

Plain soymilk or soy beverages can be formulated to have a protein content of 3 to 3.5% (Europe and Japan), 2.5 to 2.7% (United States), or less (Singapore, Hong Kong, and Malaysia), with low sugar and salt (Europe) or loaded with sugar (North America).[5,15] The flavoring added is up to the creativity of the food formulator and requirements of the local market and consumer taste. Soymilk comes in plain, vanilla, chocolate, strawberry, or any other fruit flavor and in whole, low-fat, or nonfat form. Soymilk products are stabilized before packaging. Pasteurization is used for the chilled market with a limited shelf-life of 8 to 12 weeks; sterilization or, better, UHT is used for longer shelf-life. Soymilk also appears as instant powder, creamer, coffee drink, and nutritional drink.

Tofu or soybean curd is made directly from soy base, even without intermediary cooling. In the traditional process, nigari, a sea-salt-based calcium salt blend, induces curd formation during subsequent heating. The calcium–soy protein complexes destabilize the colloid-type protein gel, producing insoluble protein precipitates. In a food technology approach, the same effect can be obtained using a soluble calcium salt such as calcium sulfate (gypsum) or simply under more acidic conditions, always combined with heat. A special case is the production of so-called silken tofu. When subjected to heat, glucono-δ-lactone added to the soy base decomposes, releasing gluconic acid that curdles the soy base in the packed product to form a very soft, silken structure and mouthfeel. Tofu or soybean curd is the bland-tasting base for a multitude of vegetarian products such as cheese and sausages but is also a versatile ingredient for cooking savory as well as sweet products.

Puddings, custards, desserts, shakes, smoothies, and juice blends are easy to make with a concentrated soy base extract. Desserts often are formulated with starch, sugars, and flavors and obtain their final structure after heating in the packaged form. The introduction of juice blends has been very successful in South America; strongly flavored products also are very attractive for markets in Africa and southern Europe.

Soy yogurt is a creamy, pudding-like food made from soy base, yogurt cultures, various sweeteners, and fruits. A higher protein level is necessary for soy yogurt to obtain the desired curd-like structure of drinkable or spoonable yogurt. It comes in many flavors and may contain active cultures. Soy base does not contain lactose, the fermentable sugar in cow's milk. Adding lactose is only an option when blends with dairy products are acceptable. For fermented soy yogurt, as such, the use of soy-adapted strains of lactobacilli and streptococci is recommended; also, cheese-type fermentation can eliminate any remaining traces of beany flavor. Concentrating the soy base with ultrafiltration to gain more consistency in the final yogurt is an option; however, it is less expensive to start with more concentrated soy base (e.g., from a process that includes okara washing).

Soy frozen desserts come in an array of traditional flavors. Frozen desserts can be made from soymilk, soy yogurt, and tofu or soy proteins, in combination with soy oil or specially formulated fats. Soy frozen desserts are available as spoonable products, ice cream bars, and ice cream sandwiches.

Soy cheese alternatives mimic traditional cheese varieties such as mozzarella, cheddar, pepper jack, parmesan, jalapeño, and cream cheese. Soy cheese may be made from tofu or soymilk or a combination of soymilk with tofu and soy protein isolates. Recent developments have allowed the production of properly fermented and ripened soy cheese in a variety of hard and soft forms and in several flavors.

Soy base can also be dried, giving soymilk powder. Oxidation and off-taste development are the primary risks when making soymilk powders. Soymilk powders are used for instant drinks as well as ingredients for food products.

References

1. Anon., INTSOY develops new techniques for commercial soymilk processing, *INTSOY Newsletter*, No. 36, University of Illinois, Urbana-Champaign, 1987.
2. Debruyne, I., unpublished data.
3. Gavin M. and Wettstein, A., *Soymilk and Other Soy Products: From the Traditional Method of Production to the New Manufacturing Processes*, Buhler, Uzwil, Switzerland, 1990.
4. Holt, S., *The Soy Revolution: The Food for the Next Millennium*, M. Evans & Co., New York, 1998.
5. *Japan Agricultural Standards (JAS) for Soymilk Products*, Notification No. 1800, November 16, 1981; Notification No. 1281, June 1, 1984; Notification No. 1482, October 5, 1985, Ministry of Agriculture, Forestry and Fisheries, Tokyo.
6. Liu, K., *Soybeans: Chemistry, Technology, and Utilization*, Chapman & Hall, New York, 1997, pp. 412–438.
7. Liu, K., *Soybeans as Functional Foods and Ingredients*, AOCS Press, Champaign, IL, 2004.
8. Maeda, H. et al., Emulsifier, Emulsion Composition, and Powdery Composition, Japanese Patent (application) 0221879, 1992.
9. MicroSoy Corporation, Jefferson, IA (http://www.microsoyflakes.com/index.htm).
10. ProSoya, Inc., Ottawa, Canada (http://www.prosoya.com).
11. Smith, A.K. and Circle, S.J., *Soybeans: Chemistry and Technology*. Vol. 1. *Proteins*, AVI Publishing, Westport, CT, 1980.
12. Soyatech and Arthur D. Little e-Sensory Perceptions™, *Sensory Benchmarking: The U.S. Soymilk Market*, Soyatech, Inc., Bal Harbor, ME, 2001 (http://www.soyatech.com/company/pr/flavor/ldml).

13. STS, *Soymilk in Brief*, Soya Technology Systems, Singapore, 1986.
14. Anon., Japan Cellfoods announces breakthrough in soybean processing at press conference, *The Soy Daily*, July 19, 2002 (http://thesoydailyclub.com/IFT62002/cellfoodsPC7152002.asp).
15. *Voluntary Standards for the Composition and Labeling of Soymilk in the United States*, Soyfoods Association of North America, Washington, D.C., 1996.

8

Meat Alternatives

Brad Strahm

CONTENTS

Introduction

Recent developments in nutrition, dietary trends, production agriculture, and world markets have resulted in an increased interest in textured vegetable proteins. The approval of a health claim for soy-based foods by the U.S. Food and Drug Administration (FDA) in the United States, as well as dietary trends toward higher protein diets, has produced greater interest in texturized soy-based products. On the other hand, the advent of genetically modified soy varieties, which greatly assist farmers in the production of soybeans, has raised product safety questions and generated interest in non-soy proteins such as wheat gluten, pea proteins, and bean proteins. In general, the food market worldwide has become more complex and demand for more sophisticated meat alternatives has grown.

Meat alternatives are a broad category of products that can include items having a wide array of textures that are manufactured utilizing various processing technologies. For the purpose of our discussion in this text, meat alternatives can be described as food items that wholly or partially take the place of meat in the human diet and that have an appearance, texture, and nutritional content similar to meat products. Meat products that might be imitated by meat alternatives are products made from emulsified meat, such as hot dogs; products made from ground meat, such as sausage or ground beef patties; restructured meat products made from whole-muscle meat, such as boneless deli ham; and whole-muscle meat products.

Many meat alternatives are manufactured by first processing protein from plant sources (vegetable proteins) into a texturized product that has a texture that imitates the chewy texture of meat. As is discussed in this chapter, these intermediate products range from simple, unsophisticated, meat-like textures to complex, more sophisticated textures that very closely match the texture of meat. The basic process technologies that transform vegetable proteins into products suitable for use in meat alternative products have been available for quite some time. These processes are generally referred to as *texturization* processes and fall into two categories: protein spinning and extrusion. In addition, some meat alternative products can be made from very traditional products such as tofu; however, because tofu is the subject of another chapter, it is not discussed here.

Types of Intermediate Products

As already mentioned, meat alternatives are generally produced from an intermediate texturized vegetable protein product. These intermediate products can cover a wide range of sophistication in terms of their textural characteristics and the process employed to manufacture them. These intermediate products can be divided into the following categories: chunk and minced products, structured meat analogs, spun protein isolates, fibrous protein products, and high-moisture meat analogs.

Chunk and Minced Products

This category of products is the one originally developed by extrusion processing in the Wenger laboratory in 1961. These products are characterized by a random spongy meat-like structure that imitates the chewy texture of meat when hydrated with water. They are typically purchased in a dry form (6 to 10% moisture) in sizes that range from small flakes or granules of about 2 mm up to large "steaks" that are 12 mm thick by 80 mm wide by 120 mm long. Other shapes that are typically available are cubes (6 to 20 mm) and noodles. The products may be colored to mimic a particular type of meat. For example, caramel-colored pieces might be used to mimic cooked red meat or red-colored pieces might be used to mimic a cured meat product. As discussed below, these products are hydrated during their transformation into a meat extender or meat alternative. This hydration process typically requires 5 to 15 minutes in room-temperature water and results in the dry product soaking up 1.5 to 3 times its weight in water. These products are typically often used as meat extenders where the granules are hydrated in water and flavors, then mixed with ground meat. This serves to extend the meat with a lower cost protein source and also modifies the characteristics of the meat to improve its ability to hold fat and water, resulting in a juicier, more palatable cooked meat product.

The smaller products from the meat alternatives category are used to manufacture products that imitate meat products made from ground meat. In this case, the granules are hydrated in a mixture of water and flavors and then mixed with binders. These binders are often soluble proteins from sources such as eggs, milk, or soy but can also include other binders such as starch and carboxymethylcellulose (CMC). This mixture is then formed by pressing it into a shape such as a patty or stuffed into a casing like sausage. The formed shapes are then cooked to set the binders so the shape is retained. These meat alternative products are then frozen and are commonly available in most U.S. supermarkets as vegetarian meat alternatives. These products can also be used to substitute for ground meat in prepared foods — for example, as a substitute for ground beef in a Southwest dish or in canned chili. The larger sized products in this category are often used as basic meat analogs. They are soaked in water and flavors and then packaged or included in a ready-meal product. An example would be a 20-mm cube of product that is used to imitate chunks of chicken breast in a ready-meal stir-fry product.

These products are typically produced via the extrusion cooking process from solvent-extracted soy flour or from soy protein concentrates. Using these materials, either a single-screw or twin-screw extruder is used. After extrusion, the product may be wet milled (discussed later) to produce sizes smaller than about 8 mm and then dried to less than 8 to 10% moisture to make it shelf stable. Before packaging, the product is sifted into size ranges to remove fines and produce products of uniform size distribution. Products are rarely flavored during the extrusion process due to the flavors being flashed off as the product exits the extruder die. If flavored products are desired, they are manufactured by externally coating a mixture of oil and flavor after drying.

For product developers and consumers who are concerned about genetically modified soybeans or who wish to make products using an identity-preserved (IP) program, an alternative raw material is mechanically expressed soy flour. Mechanically expressed soy flour is made by dehulling the beans, extruding the soy cotyledons to free the oil, mechanically pressing the extruded material to remove a portion of the oil, and then grinding the resulting cake. Texturized products containing up to 10% residual oil have been successfully and uniformly texturized using a properly configured twin-screw extrusion system. Products made from this raw material can be utilized in much the same fashion as those produced from defatted soy flour or from soy concentrate, but they will tend to have darker colors and softer textures due to their extensive heat treatment and lower protein content.

Structured Meat Analogs

Structured meat analogs (SMAs) are a reintroduction of a product that is similar to the Uni-Tex product developed by Wenger in 1975. The product is characterized by a meat-like layered structure that imitates chunks of whole-muscle meat. Its density is much higher than the chunk-style products already discussed. Where available, it can be purchased as a dry product of 8 to 10% moisture. It can be colored to mimic the appearance of cooked red meat or cured meat. Transforming SMAs into ready-to-eat meat alternatives requires hydration in water and flavors which takes 15 to 20 minutes under boiling conditions or as much as several hours in cold water. About 1 to 2 times the dry product's weight in water is absorbed by the dry SMA. It can then be utilized in any product where a diced meat product might be used, such as ready meals or canned products. When first introduced in the mid-1970s, this product was manufactured using a relatively complex system of two single-screw extruders operating in series. The first extruder served as a precooker and the second as a forming device. Now, it has been discovered that the product can be made using one twin-screw extruder system that relies on good preconditioning, and the cooking and forming functions take place in the extruder proper. The die is designed in a very streamlined fashion to limit expansion, resulting in a dense product. This product is conveyed to a dryer where long retention times are required to remove water from the dense structure. After drying, the product may be sifted to remove any fines and is then packaged.

Spun Protein Isolates

Spun protein isolates are characterized by a very meat-like fibrous structure that closely mimics the layered and fibrous structure of meat. They are manufactured from soy protein isolates and are spun as a slurry that is forced through a die plate with very small openings into a coagulating bath. After the fibers are formed in the bath, they are neutralized, compressed, and dried. The spun fibers can then be bound together by mixing them with binders. These binders are typically soluble proteins from sources such as eggs, milk, or soy and can also include non-protein binders such as starch and CMC. This mixture is formed into the desired shape to mimic a whole-muscle meat and then cooked to set the binders so the shape is retained. After cooling, the meat alternative will typically be frozen. While very nice and high-end meat alternatives can be made by this method, it is labor intensive and has serious environmental and waste disposal problems as the result of the slurries and precipitating baths. For these reasons, it is not widely used in the meat alternative industry.

Fibrous Protein Products

Fibrous protein products are extrusion-texturized products that are characterized by long strands of fine, silky fibers that run lengthwise throughout the piece. In a very real sense, it is a more efficient, high-capacity method for producing products similar to the spun protein isolates. Fibrous soy protein products are generally available in chunks that are about 15 to 25 mm in diameter by 30 to 70 mm long. These chunks are available in a dry form (6 to 10% moisture) and contain 50 to 70% protein. Caramel coloring can be added to mimic the color of cooked red meat or red coloring to mimic the color of cured meats. Natural-colored products are very useful for producing meat alternatives that mimic poultry or fish products.

Fibrous proteins are used to manufacture products that very closely imitate whole-muscle or restructured meat products such as deli meats, chicken breasts, and fish fillets. In this case, the fibrous chunks are hydrated in a mixture of water and flavors and then mixed with binders. These binders are often soluble proteins from sources such as eggs, milk, or soy but can also include other binders such as starch and CMC. This mixture is then formed by pressing it into a shape, which can be simple, such as a patty. Alternatively, the product can be stuffed into a casing like sausage, or a very complex shape can be created by molding a piece that closely imitates the structural appearance of a half-chicken. The formed shapes are then cooked to set the binders so the shape is retained. Additional surface treatments may be added to give the appearance of skin or grill marks. These meat alternative products are often frozen and can be packaged and marketed in a variety of ways that match the same schemes that are used for the meat products they are mimicking.

Fibrous protein products are made using a cooking extruder coupled with specialized die technology to form the continuous fibrous structure. Their formation also relies on the use of specific, high-viscosity soy protein isolates, vital wheat gluten, and starch. The available products also often include sulfites at a low level which are used to aid in texturization and continuous fiber formation.

High-Moisture Meat Analogs

This final category of products is also a relatively new meat alternative product. High-moisture meat analogs (HMMAs) are designed to mimic the properties, texture, nutritional profile, flavor, and appearance of whole-muscle meat. They can be created to have moisture, protein, and fat contents similar to whole-muscle meat products, especially meat products that are very lean or have a low fat content. This product has a densely layered and somewhat fibrous structure similar to that found in whole-muscle products. It is typically available in a frozen form and is generally produced in chunks to imitate cubed meat or in shreds to imitate pulled meat. Typical uses could be in prepared ethnic foods, frozen ready-meals, and similar high-end consumer food items. HMMAs are produced using a cooking extrusion process that relies on a relatively long twin-screw extruder to mix and heat the protein mass at moisture levels of 60 to 70%. This hot wet mass is pumped by the extruder through a long cooling die where texturization occurs. In addition to texturization, the product is cooled to less than 100°C before it exits the die so no expansion occurs, resulting in a very dense product. In addition, because the product is cooled and the large loss of steam common to other types of extrusion cooked meat alternative ingredients is avoided, HMMAs are a product that can be successfully flavored by mixing the flavor ingredients with the protein material blend. After exiting the extruder, the product is perishable and must be handled just like real meat. It will have to be cooled, shaped, surface treated, and frozen. In addition, retorting, aseptic, or chemical preservation are options. Like the fibrous products, it is typically based on highly refined soy proteins (soy protein concentrate or soy protein isolate), on vital wheat gluten, or on a combinations of these materials. Starch to help with the flowability of the mass through the long cooling die and oil are also often part of the recipe.

Principles of Extrusion

The vast majority of the half-products used in the manufacture of meat alternatives are texturized via the extrusion cooking process. Extrusion cooking has been defined as the "process by which moistened, expansile, starchy, and/or proteinaceous materials are plasticized in a tube by a combination of moisture, pressure, heat, and mechanical shear. This results in elevated product temperatures within the tube, the gelatinization of starchy components, the denaturization of proteins, the stretching or restructuring of tractile components, and the exothermic expansion of the extrudate."[1] Extrusion is widely used to accomplish this restructuring of protein-based foodstuffs in the manufacture of a variety of meat alternatives and meat extenders. When mechanical and thermal energy is applied during the extrusion process, the macromolecules in the proteinaceous ingredients lose their native, organized structure and form a continuous, viscoelastic mass. The extruder barrel, screws, and die align the molecules in the direction of flow, resulting in cross-linking and texturization.

Raw Materials

The raw materials used for the production of extrusion-processed textured vegetable proteins are the most expensive and important component of the process. Several important raw material characteristics require attention. These include protein level, protein quality, oil level, fiber level, sugar level and type, and particle size. The protein level of the raw material is the most important raw material property in terms of the characteristics of the product made from that material and the process required to transform the raw material into the product. In general, raw materials with higher protein levels are more easily texturized with lower levels of energy input. In addition, higher protein levels tend to result in products having tougher and firmer textures.

In addition to the protein level, the protein quality is also very important. For soy-based raw materials, protein quality is usually measured by the protein dispersibility index (PDI)[2] or nitrogen solubility index (NSI).[3] Both the PDI and NSI tests indicate the level of protein solubility in water. These indices reflect the heat treatment history of the raw material throughout the raw material preparation process. Higher levels of heat treatment will result in lower PDI or NSI values. In general, the PDI test will give lower results than the NSI test. If the raw materials have a lower PDI, more mechanical energy is required to effectively texturize them.

The whole grains from which the protein sources are derived usually contain some level of oil. Whole soybeans, for example, contain about 20% oil. In the extrusion process, oil acts as a lubricant within the extruder barrel and interferes with the addition of mechanical energy to the product. Raw materials containing higher oil levels will require an alteration of the screw and die configuration of the extruder in order to effectively texturize them. In addition to the lubrication effect, higher residual oil levels also act to dilute the protein level. Oil levels can be measured by one of two methods. The most common method is by petroleum ether extract. This method will detect oil that is not complexed in some fashion with the carbohydrate or protein fraction of the material. An alternative method is the acid hydrolysis method, which will detect all of the oil, even that which is complexed with proteins or carbohydrates. The acid hydrolysis method will always give higher results than the ether extract method.

Also, the whole grain sources of these raw materials usually contain significant levels of fiber, usually concentrated in the hull or pericarp portion of the seed. Fiber interferes with texturization by diluting the protein level and causing discontinuities in the texturized matrix. Sugars are naturally present in the source grains. If these sugars are not removed from the raw materials, they, too, act as a diluent and lower the protein content. In addition, many of these sugars ferment in the lower gastrointestinal tract, resulting in digestion problems and flatulence. For these reasons, they are removed from some raw materials, especially those that are soy based.

In any extrusion process, the particle size of the raw materials is important. Large particles are difficult to hydrate and may require additional preconditioning or additional mechanical energy input in order to plasticize and disperse the entire particle. In some cases, very fine, floury particles are also detrimental because they tend to agglomerate in the preconditioner and then these agglomerates are difficult to redisperse in the extruder barrel.

Soy-Based Raw Materials

The most popular raw material source for the production of texturized vegetable proteins is the soybean. In recent years, the worldwide production of soybeans has grown to exceed 150 million metric tons.[4] Whole soybeans typically contain about 35 to 47% protein, 18 to 22% oil, 4 to 6% crude fiber, 4 to 6% ash, and 9 to 12% moisture. From these whole soybeans, a number of raw materials can be made for use in extrusion texturization. These raw materials include defatted soy flour, soy protein concentrate, soy protein isolate, and mechanically expressed soy flour. Defatted soy flour is made by first cleaning, cracking, and dehulling the beans, followed by conditioning and flaking. The full-fat flakes are then processed through a solvent extractor where the oil is solubilized in an organic solvent (usually hexane) and removed. The flakes are toasted or flash desolventized to remove residual solvent, then ground and sized to produce soy flour. As shown in Table 8.1, the degree of toasting has a large impact on the protein quality as measured by the PDI. Defatted soy flour to be used for the extrusion texturization process should have the characteristics shown in Table 8.2. Products made from defatted soy flour will have a flavor that is commonly associated with soy and will tend to cause flatulence problems, as evidenced by its high sugar content.

TABLE 8.1

Protein Dispersibility Indexes of Soy
Flour

Types of Soy Flour	PDI
Negligible toasting	85–95
Light toasting	70–80
Light to moderate toasting	34–45
Toasted	8–20

TABLE 8.2

Defatted Soy Flour for Extrusion
Texturization Process

Component	Value
Protein	50–55%
Oil (pet ether extract)	<1%
Crude fiber	<3.5%
Sugars	12%
PDI	50–70
Particle size	~100 mesh

TABLE 8.3

Soy Protein Concentrate

Component	Value
Protein	65–70%
Oil (pet ether extract)	<1%
Crude fiber	<4%
Sugars	1%
PDI	Very low
Particle size	100–200 mesh

Soy protein concentrate is made by further processing toasted, defatted soy flakes and through an aqueous alcohol process to remove the soluble sugars. After drying, it is ground to a powder to produce soy protein concentrate having the approximate composition shown in Table 8.3. The sugars have been removed, so products made from soy protein concentrate are easily digested and contribute less to flatulence problems. Although soy protein concentrates made by this process have a very low protein dispersibility index, the product is very easily texturized. Soy protein concentrate usually requires processing at higher moistures and slightly higher mechanical energy inputs than defatted soy flour. It is worthwhile to note that a high PDI soy protein concentrate product is available, as well. While it is useful in some of the more sophisticated products such as high-moisture meat analogs and fibrous protein products, it is not commonly used to manufacture chunk-style products because of its higher cost.

Soy protein isolate is made by processing defatted flakes through a process that selectively solubilizes the proteins. The proteins are then precipitated, conditioned, and dried to form soy protein isolates. The wide variety of soy protein isolates available on the market today have been customized for various applications, including extruded products. It is imperative that the product developer work with the ingredient supplier to ensure that the correct soy isolate is chosen for a particular application. Soy protein isolate is commonly used in more sophisticated products such as high-moisture meat analogs and fibrous protein, as discussed previously.

Mechanically expressed soy flour is made by a process much different than that used to make defatted soy products. In this process, whole soybeans are cleaned, cracked, and dehulled, then they are extrusion processed to free the oil and inactivate antinutritional factors and lipoxygenase enzymes. After extrusion, the resulting meal is immediately passed through a mechanical screw-type press where oil is expressed from the meal. The resulting meal cake is then cooled and ground to prepare it for extrusion. As with the solvent extraction process, the level of heat treatment (in this case, the pre-expelling extrusion step) will determine the PDI of the resulting material. In general, a lower temperature extrusion treatment will result in a higher PDI and higher oil content, and a higher temperature extrusion treatment will result in a lower PDI and lower oil content.

The advantages of this type of process are that it does not involve an organic solvent, natural antioxidants are retained in the meal, and the resulting flour has a sweeter, more pleasing flavor. For the best results, it has been found that mechanically defatted flour having the characteristics shown in Table 8.4 is most easily texturized. To texturize this material, dramatic adjustments in the screw and die design are required to overcome the lubricating effect of the high residual oil content and low PDI. Disadvantages of this material are that the products are less uniform in size and shape due to high oil content, are darker in color due to extensive heat treatment in preparing the raw material, and are softer in texture due to a relatively low protein content.

Wheat Gluten

The popularity of textured vegetable proteins that contain wheat proteins is rapidly increasing. As already mentioned, wheat proteins are commonly used in fibrous protein products but are also often used in the other product categories as well. The wheat kernel is composed of approximately about 12 to 16% protein, 2 to 3% oil, 2 to 3% fiber, 1 to 2% ash, and 8 to 10% moisture. The wheat milling process produces wheat flour that has most of the fat and ash portions removed. After milling into flour, wheat gluten (the protein portion of wheat flour) is separated from the starch by one of a number of processes. Wheat gluten is a protein that has very unique properties. When hydrated and mixed, it forms a very extensible, elastic structure that is responsible for the gas-holding ability of bread dough. Wheat gluten can be

TABLE 8.4

Mechanically Defatted Flour

Characteristic	Value
Protein	40–50%
Oil (pet ether extract)	10–12%
Crude fiber	3–4%
PDI	25–30
Particle size	100–200 mesh

used in combination with soy-based raw materials or in combination with wheat flour and other additives to produce a soy-free texturized product. Commercially available wheat gluten is typically 80% protein.

Other Protein Sources

In addition to soy and wheat gluten, other raw material sources have been used in the extrusion process to produce products that can be made into meat alternatives. Meat extenders have been extruded from proteins from cotton-seed, rapeseed, canola, peanuts, sesame, sunflowers, peas, and beans with varying results. No matter the protein source, it seems that the most important properties of those raw materials are the ones recited earlier: protein level, protein quality, oil level, fiber level, sugar level and type, and particle size.

Additives

Minor ingredients or chemicals are sometimes added to increase the range of raw ingredients suitable for production of a specific textured vegetable protein product. It may be important to understand the effects of these additives when attempting to fine-tune the textural qualities of a product. Increasing the pH of vegetable proteins before or during the extrusion process can aid in the texturization of the protein, especially those that for some reason have a low starting pH. Extreme increases in pH increase the solubility and decrease the textural integrity of the final product.[5] Processing at pH 8.0 may also result in the production of harmful lysinoalanines.[6] Lowering the pH has the opposite effect and decreases protein solubility, making the protein more difficult to process. Undesirable sour flavors in texturized vegetable protein products may be evident if the pH is adjusted below pH 5.0. Soy lecithin added to formulations of vegetable proteins at levels up to 0.4% tend to assist in smooth laminar flow in the extruder barrel and die, which permits production of increased-density soy products. The ability to make dense vegetable protein products is related to the higher degree of cross-linking occurring during the extrusion process.

Sulfur

Known for its ability to aid in the cleavage of disulfide bonds, sulfur assists the unraveling of long twisted protein molecules. This reaction with the protein molecules results in increased expansion and a smooth product surface and adds stability to the extrusion process. These benefits, however, are not without some undesirable side effects, including off-flavors and unpleasant aroma. Normal dosing levels for sulfur or sulfur derivatives are in the 0.01 to 0.2% range.

Calcium

Increased textural integrity and smooth product surfaces are the result of incorporating calcium chloride into a textured vegetable protein product. Dosing levels for CaCl range between 0.5 and 2.0%. With the addition of CaCl and small amounts of sulfur, soybean meal containing 7.0% fiber may be texturized, retorted for 1 hour at 110°C, and still maintain a strong meat-like texture.

Textured Vegetable Protein Production System

A typical process flow diagram for the production of textured vegetable proteins is shown in Figure 8.1. Raw materials are supplied either in bulk, as shown, or in bags. After mixing, if necessary, the raw materials are conveyed to the live bin of the extrusion portion of the system. The extrusion portion of the system includes a live bin/feeder, preconditioner, extrusion cooker, and die/knife assembly, as shown in Figure 8.2. The design of each of these components is engineered to accomplish a specific function in the process of texturizing vegetable food proteins. Within the design features, the operating conditions are adjusted to vary the texture of the finished product. The effects of each processing step on the product are specifically addressed.

Feeder

The live bin/feeder provides a means of uniformly metering the raw materials (granular or floury in nature) into the preconditioner and subsequently into the extruder. This flow of raw material must be uninterrupted and rate controllable. The live bin/feeder controls the product rate or throughput of the entire system.

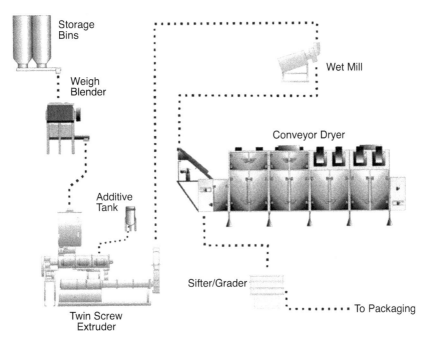

FIGURE 8.1
Process flow diagram for the production of extruded texturized vegetable proteins.

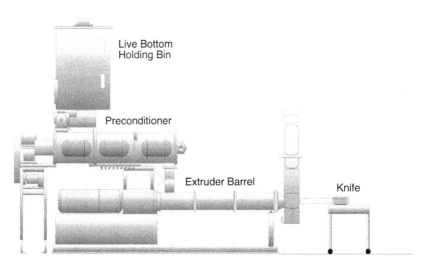

FIGURE 8.2
Extruder for manufacturing texturized vegetable proteins.

Preconditioner

Without the use of preconditioning it is difficult to make good laminar-structured textured vegetable protein. Unpreconditioned vegetable proteins have a strong tendency to expand rather than laminate due to nonuniform moisture penetration that does not allow uniform alignment of protein molecules. Uniform and complete moisture penetration of raw ingredients significantly improves the stability of the extruder and final product quality. In addition, by completely plasticizing the raw material particles prior to their introduction into the extruder barrel, extruder wear caused by the abrasive raw material particles is greatly reduced. In the atmospheric preconditioner, moisture is uniformly applied in the form of water or steam to achieve a moisture content of 10 to 25%. Water is introduced from the top of the preconditioner by means of spray nozzles to atomize the water stream, thereby reducing the mixing load on the preconditioner. Steam is added from the bottom of the preconditioner. The steam-supply plumbing must be designed to maintain a continuous flow of dry, saturated steam. If the steam added to the preconditioner contains pockets of condensate, an unstable extrusion process will result due to rapidly varying moisture content. The preconditioner is vented to release excess steam and undesirable volatile flavor components found in the raw vegetable protein. Most preconditioners contain one or two mixing and conveying elements consisting of rotating shafts with radially attached pitched paddles. Atmospheric or pressurized chambers may be used in the preconditioning step. Pressurized preconditioners can achieve higher discharge temperatures but have the disadvantages of potential nutrient destruction[7] and elevated purchasing and operating costs.

Extruder

The preconditioner discharges directly in to the extruder assembly consisting of the barrel and screw configuration. Here, the major transformation of the raw preconditioned vegetable proteins occurs which ultimately determines the final product structure. Extruders used for the manufacture of textured vegetable proteins are either single screw or twin screw in design. In both cases, the impact on final product texture is produced by the screw and barrel profile, screw speed, processing conditions (temperature, moisture), raw material characteristics, and die selection. The successful production of meat extenders from defatted vegetable proteins on single-screw extruders is clearly dependent on appropriate PDI or NSI values and consistently defatted materials. Twin-screw extruders, on the other hand, are much more flexible and can be configured to produce a much wider range of products with a much wider range of raw materials.

Recent analysis of the production costs of single-screw vs. twin-screw extruders has shown that, although the twin-screw extruder has higher capital investment costs, its long-term operating costs are similar to the single-screw. Coupling this effect with greater uptime and process stability results in the twin-screw extruder being the more cost-efficient option for most producers.[8]

The initial section of the extruder barrel is designed to act as a feeding or metering zone to simply convey the preconditioned vegetable protein away from the inlet zone of the barrel and into the extruder. The material then enters a processing zone, where the amorphous, free-flowing vegetable protein is worked into a colloidal dough. The compression ratio of the screw profile is increased in this stage to assist in blending water or steam with the raw material. The temperature of the moist proteinaceous dough is rapidly elevated in the final 2 to 5 seconds of dwell time within the extruder barrel. Most of the temperature rise in the extruder barrel is due to mechanical energy being dissipated through the rotating screw and may be assisted by the direct injection of steam or from external thermal energy sources. The screw profile may be altered by choosing screw elements of different flight pitch or via interrupted flighting, or by adding mixing lobes configured to convey in either a reverse or forward direction. All of these factors affect the conveying of plasticized material down the screw channel and therefore the amount of mechanical energy added via the screw.

In the extruder barrel, hydration and heating cause unraveling of the long twisted protein molecules in vegetable proteins. In the extrusion process, these molecules align themselves along the streamline flows of the screws and dies. The extent of cross-linking seems to be a function of time, temperature, and moisture history.

The proper exposure to shearing action as the protein molecules align themselves for cross-linking during the extrusion process is important. The test data shown in Figure 8.3 demonstrate that, although adequate shear rates are necessary to enhance the cross-linking effects of protein molecules, over-shearing after the cross-linking has begun to occur may disrupt the layered structure of the protein molecules, resulting in decreased water-holding capacity.

Figure 8.3 shows the relationship between the water-absorbing ability of a texturized soy concentrate and the level of mechanical energy input. As indicated by these data, the optimum water absorption occurs at about 390 kJ of mechanical energy input per kilogram of material processed. In this case, water absorption is expressed as the amount of water absorbed divided by the initial dry sample weight.

The effect of moisture on the rheology of soy protein doughs is a very important processing variable that may be manipulated. While decreased moisture levels can increase the amount of mechanical energy input, increased moisture levels increase mobility of the chemically reactive protein molecules, thus increasing the potential for reactive sites to come in close proximity, facilitating cross-linking, and resulting in increased extrudate density.

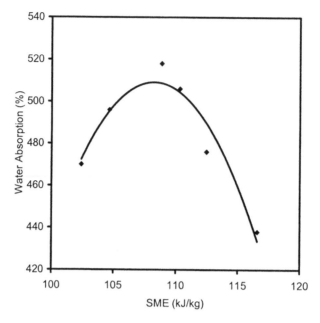

FIGURE 8.3
Effect of mechanical energy added during extrusion on the water absorption capacity of texturized vegetable protein.

Die

The extrusion chamber must be capped with a final die, which serves two major functions. First, the die offers restriction to product flow, thereby causing the extruder to develop the required pressure and shear, and, second, the final die shapes the product. The plasticized material is extruded through the die openings, and expansion occurs as the product is released to ambient pressures. Final product density has consequently been shown to correlate with the extrusion temperature and moisture levels. Dies for texturized vegetable proteins are usually one of two types: face dies or peripheral dies. The openings of face dies are located such that the extrudate exits the die in the same axial direction as the overall flow through the extruder. In the case of peripheral dies, the extrudate exits the die at right angles to the direction of the overall flow through the extruder. The choice of which type of die to utilize for a particular product and process depends entirely on the nature of the product and the raw materials used to produce it. In cases where high die restriction is desirable to increase mechanical energy input, but large amounts of open area are required in the final die for proper product shaping, a Venturi die concept may be used. Regardless of their type, dies for texturized vegetable proteins should have smooth, streamlined flows that do not disrupt or cause shearing effects to the already laminated and cross-linked protein molecules.

Post-Extrusion Processing

The textured vegetable protein is discharged from the extrusion system and is carried either by belt conveyor or pneumatic conveying line directly to the dryer or to a wet milling device. The wet milling device is usually either a hammer mill or an Urschel Comitrol cutting device. The hammer mill device will tend to give a more flaked appearance, while the Urschel Comitrol will tend to result in a sliced effect. After wet milling, the product is then conveyed to the dryer.

Drying

The dryer used for textured vegetable products is usually a conveyor-style dryer, where the drying air is heated with either steam or a combustion burner. There is some concern in the industry that, in direct-fired dryers where the products of combustion are mixed with air circulated through the product, the combustion products can combine with the proteins to produce carcinogenic substances. To avoid this in dryers that are heated with a combustion burner, indirectly fired systems are available that heat the process air via a heat exchanger, and the products of combustion are discharged without contacting the product. In the horizontal conveyor dryer, the product is spread on a belt that moves through zones where heated air is passed through the product. After the air is circulated through the product, a portion of it is exhausted to carry away the water removed from the product, and the remainder is mixed with fresh incoming air, reheated, and passed through the product again. The units can be single or multiple pass in design, depending on the configuration required to fit in the factory and adequately dry the product. The horizontal conveyor dryer provides for excellent control of retention time and results in uniform drying.

A number of product factors determine how the product dries. The moisture content, size, shape, and density of the incoming product can all alter its characteristic drying curve. Temperature, time, bed depth, and air velocity are all controlled within the dryer to accomplish both complete and uniform drying. Drying curves are used to describe how a product dries. An example of a drying curve for texturized soy flour chunks dried at 100°C air temperature is shown in Figure 8.4. This figure demonstrates how the product moisture and temperature change during the drying process. Note that the product moisture changes more rapidly during the first portion of the drying process, and then the product loses moisture more slowly as drying progresses. Also note that the product temperature is about 80°C as it is supplied to the dryer from the extruder and, as a result of evaporative cooling, the product temperature is lowered during the first phase of drying to a minimum of about 69°C. After the surface moisture is removed, the product temperature begins to rise. This is a result of heat being added to

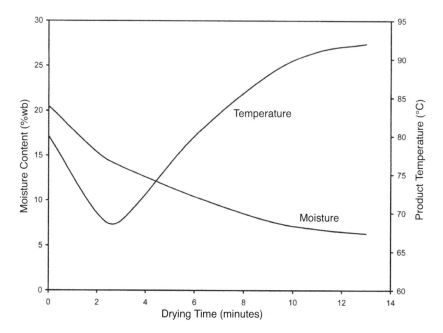

FIGURE 8.4
Drying curve for texturized soy flour chunks dried at 100°C.

the product more quickly than water can be evaporated due to restrictions on water movement within the product. The moisture content corresponding to the point at which the temperature begins to rise is the critical moisture content, in this case about 14% moisture. For a target final moisture content of 10%, the required drying time of 6.5 minutes can be read from Figure 8.4.

Examples of drying curves for three different products are shown in Figure 8.5. Note that each of these products has a different starting moisture as a result of their processing requirements in the extruder. In addition, due to differing product structure, each of these products releases its moisture at a different rate, resulting in different drying times. For the examples shown here, the soy flour chunk and wet-milled concentrate both require about 5 minutes of drying time to reach 10% moisture. Even though the wet-milled concentrate starts at a significantly higher moisture, its product size and shape allow it to release water more quickly compared to the soy flour chunk. The dense meat analog requires about 19 minutes of drying time, longer than the other products due to its higher incoming moisture and the slow release of water because of its dense structure.

After drying, the product is cooled, fines are removed, and it is segregated into appropriate size ranges and sent to holding bins prior to packaging. As mentioned earlier, this stage of the processing can also include an external coating step where flavor and oil are coated on the exterior surface of particles.

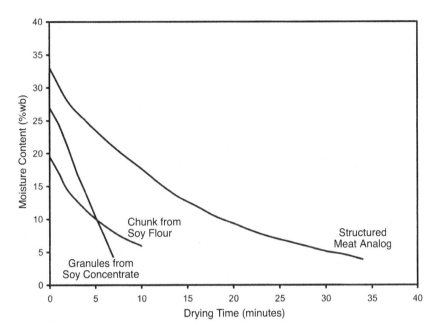

FIGURE 8.5
Example drying curves for various types of texturized vegetable protein products.

Conclusion

The changing face of consumer desires and technology in the food industry is driving continued interest in meat alternative products. These products serve to fill the product category of vegetarian, high-protein, low-fat foods. Some products are quite simple and can be produced at lower cost than meat itself, while others are quite sophisticated and their costs can approach or even exceed those of meat products. Certainly, the overall result is that consumers have much more choice in meat alternatives today, and it appears that this will continue to be the trend into the future.

References

1. Smith, O.B., Textured Vegetable Proteins, paper presented at the World Soybean Research Conference, University of Illinois, August 3–8, 1975.
2. AACC Method 46-24, Protein Dispersibility Index, rev. October 26, 1994, in *Approved Methods of the American Association of Cereal Chemists*, 9th ed., American Association of Cereal Chemists, St. Paul, MN, 1995.

3. AACC Method 46-23, Nitrogen Solubility Index, rev. October 26, 1994, in *Approved Methods of the American Association of Cereal Chemists*, 9th ed., American Association of Cereal Chemists, St. Paul, MN, 1995.
4. Anon., Production of oilseeds forecast to rise, *Financial Times, London Edition*, September 23, 1999.
5. Boison, G., Taranto, M.V., and Cheryan, M., Extrusion of defatted soy flour–hydrocolloid mixtures: effect of operating parameters on selected textural and physical properties, *J. Food Technol.*, 18, 719, 1983.
6. Simonsky, R.W. and Stanley, D.W., Texture–structure relationships in textured soy protein. V. Influence of pH and protein acylation on extrusion texturization meat analogs, meat extenders, *Can. Inst. Food Sci. Technol. J.*, 15, 294, 1982.
7. DeMulenaere, H.J.H. and Buzzard, J.L., Cooker extruders in service of world feeding, *Food Technol,*, 23, 345–351, 1969.
8. Plattner, B., *Texturized Vegetable Protein Short Course*, Texas A&M University, College Station, 2000.

9

Textured Soy Protein Utilization in Meat and Meat Analog Products

M.W. Orcutt, M.K. McMindes, H. Chu, I.N. Mueller, B. Bater, and A.L. Orcutt

CONTENTS

Introduction

Textured vegetable proteins utilized for the manufacture of various meat and meat analog products represent vast improvements over similar products of 30 and even 10 years ago. Though these improved textured protein products outwardly appear the same as products manufactured a generation ago, they are the result of research and development efforts challenged to satisfy consumer demands regarding both texture and flavor. Textured proteins designed for use in food products are available in a cornucopia of sizes, colors, textures, and protein concentrations that are utilized to provide specific characteristics to various food products. Traditionally, soy flour and soy protein concentrates have been the principal source materials for the majority of the commercial textured protein ingredients utilized in the food industry; however, products comprised of wheat gluten, soy protein, starches, and other ingredients provide an essentially endless array of textured vegetable protein ingredients that can be utilized to create both unique and interesting meat and meat analog products of specific color, texture, flavor, etc.

Textured soy proteins are common yet important ingredients of many meat and meat analog products. After hydration, textured vegetable protein materials of the various sizes, densities, and compositions contribute both meat-like appearance and texture to food products while providing an excellent source of high-quality protein similar to that of lean meat. Though inherently bland with regard to flavor, textured soy proteins are excellent absorbers of both artificial and natural flavors. Additionally, textured soy protein ingredients are easily colored using caramel colors, spice pigments, and seasoning ingredients.

The combined characteristics offered by the various commercial vegetable textured proteins can be designed into a meat product to impart a nutritional profile and an eating experience similar to those of an all-meat product. Textured soy proteins are used throughout the world primarily to replace meat for the purpose of formulation cost reduction; however, these textured ingredients contribute other important attributes to the foods into which they have been formulated. Textured soy proteins hold water very tenaciously. This attribute allows meat products made with textured soy material to have greater moisture retention during cooking, reheating, holding in readiness for serving, and through freezing and thawing compared to similar meat products made without such ingredients.[1] Judicious use of textured soy protein ingredients in meat formulations allows the manufacture of both frozen raw and precooked meat products possessing better eating quality attributes compared to similar products made without textured soy protein products. Additionally, inclusion of textured soy protein ingredients in coarse ground products such as meat patties and meatballs disrupts the oriented collagen and myofibrillar matrices established during product forming, thus reducing dimensional distortion caused by the heat-induced shortening of these meat protein matrices that occurs during cooking. Thus, the addition of textured soy protein products to meat patties and meatballs results in rounder but flatter patties and more spherically shaped meatballs. Moreover, cook yields are higher for meat product formulations containing textured vegetable ingredients.[1]

Textured Soy Protein Products

Textured protein products utilized in both the meat and vegetarian meat analog industries include primarily textured soy flours, textured soy protein concentrates, and textured soy protein materials comprised of various blends of isolated soy protein, soy protein concentrate, soy flour, wheat flour, vital wheat gluten, rice flour, soy fiber, assorted starches, etc. The various textured vegetable protein pieces are utilized primarily as dry ingredients, becoming hydrated either prior to addition to food formulations or during preparation steps such as blending or chopping of the food material. Some textured ingredients are manufactured as fully hydrated ingredients that are either frozen immediately following preparation for shipping and handling or utilized immediately following their preparation. Such textured ingredients can be prepared via spinning, chopping, or cold extrusion.

Textured soy protein products are available in a variety of textures, sizes, colors, and flavors. Extrusion processes can be altered to provide textured soy particles of different sizes, shapes, and densities.[2] Generally, as would be expected, low-density textured soy protein particles hydrate rapidly and usually contribute softer textures to foods; denser particles tend to hydrate more slowly and impart firmer textures to foods. Data presented in Table 9.1 demonstrate the relationships among particle density, rate of hydration, and ultimate cooked water-holding capacity for textured soy protein concentrate ingredients of varying densities. Throughout the soy protein industry, the most common forms of the various textured soy protein products used in the manufacture of meat and meat analog products are flakes, crumbles, and granules. Flakes are the least dense; granules tend to be the most dense compared to other textured particles, and crumbles usually have densities between those of flakes and granules (Table 9.1). In addition to flakes, crumbles, and granules, nomenclature commonly used in the food industry to describe both texture and shape includes chunks, strips, minces, and shreds. This common terminology is only somewhat well understood throughout both the soy ingredients and meat industries. These terms typically refer to shapes rather than specific textures or product densities; the textured pieces may possess textures anywhere between those of flakes and granules. Textured materials referred to as shreds may have either an amorphous texture or a very defined fibrous texture comprised of discreet fibers oriented in a given plane. The term *mince* or *minced* refers to both the density and shape of textured vegetable protein pieces, and such textured particles are considered granular in texture.

TABLE 9.1

Water-Holding Capacities and Cooked Texture of Textured Soy Protein
Concentrate Flake, Crumble, and Granule Particles

Type of Textured Soy Protein Concentrate	Water-Holding Capacity			Kramer Shear Force (g)[b,d]
	3-Minute Hydration[a,b]	60-Minute Hydration[a,b]	Cooked[c]	
Flake (1/4-in. diameter)	3.0:1*	3.7:1*	4.4:1	59,806***
Crumble (1/4-in. diameter)	1.9:1**	3.7:1*	4.4:1	69,295**
Granule (1/4-in. diameter)	0.7:1***	1.9:1**	3.9:1	72,992*

[a] Dry textured soy protein ingredients were hydrated utilizing five parts water to one part dry textured soy protein concentrate product. Particles were hydrated for 3 and 60 minutes prior to measurement of water retention. Following hydration, excess water was pressed from the hydrated soy material by placing a 5-kg weight on hydrated material held over a number 30 mesh screen.

[b] Means sharing a common superscript are not different ($p > 0.05$).

[c] Hydration following cooking was determined by placing the hydrated textured soy protein concentrate material in 307 × 109 tuna cans and cooking the sealed cans for 30 minutes in a hotwater bath maintained at 100°C. After cooking, excess water was removed from the textured soy protein concentrate material by pressing the hydrated material with a 5-kg weight over a number 30 mesh screen.

[d] A Texture Technologies XT2 texture analyzer fitted with a 10-blade Kramer shear head was used to shear 120 g of hydrated textured soy protein concentrate material. Textured soy protein was hydrated 3:1 with water (weight-by-weight basis) prior to sealing and cooking in a 100°C water bath for 90 minutes.

Sizes commonly used in meat products, range from particles as small as 1/16 in. (1.5 mm) in diameter to chunks as large as 1/2 in. (12 mm) in diameter. Larger pieces are available but are less common than textured vegetable proteins of less than 1/2-in. (12-mm) diameter. Generally, as would be expected, larger textured particles tend to be more difficult to hydrate. This characteristic may limit their acceptability for use in some product applications as manufacturing constraints of time and equipment may preclude their selection as ingredients in certain food products even though from the standpoints of texture, appearance, etc. the large pieces allow preparation of the best product.

In addition to particle texture and size, textured vegetable protein products are available precolored; these particles can be red or red/brown, taking on the appearance of bacon or corned beef, or various shades of brown to appear as cooked beef or pork. Canned corned beef sold in the Philippines contains, as a common ingredient, either dark red textured soy flour or dark red textured soy protein concentrate. After hydration and cooking, these textured particles have the red–pink color of cured beef, and in this product it is difficult to distinguish the textured soy material from the meat component. Textured soy protein products colored to resemble salmon and tuna are also available. The texture, size, color, and shape of the various textured soy products can be customized for almost any application.

Vitamin and mineral fortification is an additional custom option for consideration. Textured soy protein ingredient fortification is often part of school food service programs throughout the United States and Canada. Fortification requirements for the various textured protein products would be uniform within any given country; however, fortification requirements vary slightly among countries.

Soy Grits

The most common forms of soy protein used in meat and meat analog formulations are soy grits, textured soy flours, and textured soy protein concentrates. Soy grits, made from coarsely ground defatted flakes that may or may not be toasted, are used in the United States in some sausage products. To some extent, inclusion of soy grits in processed meats formulations (primarily sausage products) is regional in nature. For example, in the southeastern and southern United States, soy grits are a common ingredient of some coarse-ground cooked sausage products. Additionally, soy grits commonly are used in both meat and poultry patty formulations. Soy grits are not texturized proteins, but they are utilized widely in the meat industry to provide both reduced formulation cost as well as specific textures to meat products. Soy grit ingredients are available as coarse, medium, and fine grits and as untoasted, lightly toasted, and fully toasted.[3] Soy grits are one of the least expensive forms of soy protein materials utilized to impart texture to meat foods. Inclusion of soy grits in meat product formulations at high concentrations may impart both a gritty texture and a pronounced bean or cereal flavor note to meat products; however, excellent meat products can be prepared with judicious use of soy grits in the formulations.

Textured Soy Flour

Most textured soy flours are produced via thermoplastic extrusion of defatted soy flour. As described previously in the general overview of textured vegetable protein characteristics, textured soy flour can be prepared so as to produce texturized particles of various sizes, shapes, colors, textures, and flavors. Textured soy flours are typically somewhat less functional with regard to water-holding capabilities compared to other textured vegetable protein ingredients. Additionally, textured soy flours, when compared to similar textured soy protein concentrate particles, possess softer textures after hydration, as shown in Table 9.2; this is related in part to their relatively

TABLE 9.2

Texture Comparison

Treatment[a]	Beef Patty Kramer Shear Force (g)[b]
Textured soy protein concentrate (1/4-in. [6-mm] flake)	19,956[*]
Textured soy flour (1/4-in. [6-mm] flake)	13,159[**]

[a] All beef patties were formulated to contain 30% fat. Beef patties containing soy protein material were formulated to contain 30% hydrated soy material; both the textured soy flour and textured soy protein concentrate were hydrated with three parts water to each part dry textured soy material (weight-by-weight basis). Hydrated textured soy protein replaced lean meat in the formulation. Total formulation fat content was reduced form 30 to 28% as a consequence of the meat replacement.

[b] Shear values sharing a common superscript are not different ($p > 0.05$).

high concentration of soluble sugars diluting the textured particle insoluble fiber and protein components.

The acronyms TVP and TSP are used extensively throughout the food industry in reference to textured soy flour products and are often used to describe all textured soy protein products. It should be noted, however, that these acronyms are registered trademarks; TVP® is registered to Archer Daniels Midland (ADM) and represents their textured soy flour products; TSP® is registered to Legacy Foods LLC (formerly PMS Foods LP) and represents their line of textured soy flour ingredients.[4,5]

When preparing meat formulations containing textured soy flours, the textured soy flours are hydrated using a ratio of 2 to 2.5 parts water to each part (weight-by-weight basis) of the textured soy flour material. Textured soy flours hydrated 2:1 contain approximately 16.5% protein; this is equivalent to the protein content of many lean meat ingredients utilized throughout the world for the manufacture of processed meat products. Table 9.3 provides a reference of recommended textured soy protein ingredient hydration as well as expected total protein contents of the various textured ingredients after typical or recommended hydration. The open texture of many of the soy flours used for the manufacture of various meat products allows the entrapment of both moisture and liquid meat fats released from other ingredients during cooking such that the eating quality from the aspects of texture, mouthfeel, and flavor is preserved when textured soy protein products are included in meat product formulations. It should be noted that textured soy flours usually possess more pronounced bean-like and cereal flavors compared to textured soy protein concentrates, textured isolated soy proteins, and textured products comprised of soy protein and other ingredients. The additional processing associated with the manufacture of soy protein concentrate and isolated soy protein products removes the water-soluble soybean components responsible for the bean and cereal flavors. Moreover, textured soy flours contain as much as 4% sucrose and thus tend to be sweeter compared to other textured vegetable protein ingredients.

TABLE 9.3

Textured Soy Protein Products

Textured Soy Protein Product	Protein Content (%)	Typical Hydration	Hydrated Protein Content (%)
Textured soy flour	50	1:2	16.50
Textured soy concentrate	65	1:3	16.25
Textured soy isolates	85–90	1:3	21.25
		1:4	17.00

Textured Soy Protein Concentrates

Textured soy protein concentrates can be custom manufactured as described for the textured soy flours, and essentially any color, particle or piece size, particle density, shape, flavor, and fortification can be provided. Textured soy concentrates offer functionality and flavor profile improvements over those of textured soy flour. Textured soy protein concentrates typically possess higher water-holding capacities and firmer hydrated textures compared to textured soy flours of similar size, shape, and texture (density). These characteristics were demonstrated earlier by the data provided in Table 9.2. Additionally, textured soy protein concentrates are considered to have a blander flavor than textured soy flour, as the carbohydrates soluble in polar solvents are removed during a water or alcohol wash during preparation of the soy concentrate base material used to manufacture textured soy protein concentrates. Moreover, textured soy protein concentrates tend to be less sweet compared to textured soy flours due to the removal of the soluble sugars. Textured soy protein concentrates are usually hydrated with between 2.5 and 3 parts water for each part dry ingredient (weight-by-weight basis) in a given meat or meat analog formulation; after being subjected to a heat treatment, textured soy protein concentrate ingredients would be expected to retain added water at a level of approximately four times their dry weight, as shown earlier in Table 9.1.

Textured Isolated Protein Products

Textured isolated soy proteins are not common food product ingredients. Although it is possible to manufacture excellent textured pieces utilizing only isolated soy proteins, these textured pieces may possess a harder texture compared to similar textured soy protein concentrate or textured soy flour particles. The primary reason textured pieces produced solely from isolated soy protein are not common is related to their cost. These dry textured materials are valued more than other textured soy ingredients yet do not have substantially improved water-binding capabilities compared to textured soy concentrates and other textured vegetable protein products. Moreover, they rarely offer unique characteristics with regard to texture, size, color, etc. to differentiate them from other less-expensive textured protein products.

Another way to provide a textured isolated soy protein for use in meat and meat analog products is to produce a textured hydrated gel granule at the food processing plant utilizing a highly functional isolated soy protein powder. In this patented process, 2.5 to 3.5 parts water are added to each part of an isolated soy protein characterized as having excellent gel strength.[6-8] To prepare a hydrated soy isolate granule, the water and isolated soy protein should be chopped using a bowl chopper or blended in a paddle or rubber blender until a strong gel resembling a cottage cheese curd is formed. The hydrated isolated soy protein particulate texture can be modified by using more or less hydration water. It is important to keep isolated soy protein gels as cold as possible to maintain a firm gel throughout processing of the food product. Cold gel temperatures are also necessary to prevent microbial growth after the gel has been added to the meat product. Chilling via carbon dioxide snow can be utilized to help maintain gelled particle integrity throughout blending, grinding, and forming. The isolated soy protein gel can be frozen, tempered, and coarse ground prior to adding the ground soy particle to a cold meat mixture. It is best to prepare cold isolated soy protein gels the day prior to use and to hold the gels at least overnight to allow formation of firm gel matrices. Firmer or stronger gels can be prepared by hydrating the isolated soy protein with hot water during chopping or blending; the addition of salt will further increase the gel strength. Salt should be added after hydration (via chopping or blending) of the isolated soy protein. For most applications, the hot gels would require chilling prior to incorporation into meat products. Particle color can be altered using natural colorants such as caramel coloring, carmine, and paprika, as well as various aqua and oleo resins of herbs and spices that are known to contribute color to foods.

Textured Vegetable Protein Blends

Textured soy protein products possessing unique flavor, functional, and textural characteristics can be prepared by combining various plant protein ingredients, starches, and flours prior to extrusion. Textured plant protein products comprised of blends of wheat gluten and soy protein have unique appearances and textures compared to textured plant protein products comprised of wheat gluten, soy protein concentrate, or soy flour alone. Starches and flours are common ingredients of textured vegetable protein products comprised primarily of soy and wheat proteins. Wheat gluten provides unique oriented structure to the textured pieces as well as an interesting and desirable chewy texture. Soy protein adds a unique texture profile and improves the overall protein quality of the textured vegetable protein product. Starches and flours act to soften the textured pieces. Adjusting the levels of each ingredient permits the design of textured plant protein pieces or particles possessing specific characteristics such as hydrated firmness or chewiness.

Meat Coatings

An international concern for personal health has provided new opportunities for the use of soy proteins in meat products. Various soy proteins can be the primary ingredients of predust, batter, and breading formulations, replacing carbohydrate ingredients in the batter and bread coatings. Breaded meat products containing 30% of a traditional breading would provide roughly 15 g of total carbohydrate per 100-g serving and 12 to 15 g of non-fiber carbohydrate. A similar product breaded with batters and breadings containing soy protein would provide dramatically reduced non-fiber carbohydrates per 100-g serving. Such utilization of soy protein ingredients or breading materials formulated to contain significant quantities of a soy protein ingredient as coatings for fried foods is a relatively new product development concept. In such cases, a textured particle containing a soy protein ingredient would replace traditional cracker, Japanese-style, or American crumb breading. Moreover, soy proteins may be important in the design of new fried breaded foods of the future, as soy protein ingredients may enhance the holding time of hot fried breaded products as well as improve the eating characteristics of oven-baked and microwave-prepared breaded meat and meat analog products.

Regulations Governing the Use and Labeling of Products Containing Textured Vegetable Protein Products

United States

The U.S. Department of Agriculture (USDA) directs the labeling of meat products produced and sold in the United States. A meat product, from the standpoint of sale in the United States is considered to be any food product containing more than 2% cooked meat or more than 3% raw meat.[9] One of the most common uses for textured soy protein materials in meat products is the substitution of a portion of the meat with hydrated textured soy protein concentrate or textured soy flour in both poultry and beef patty formulations. Products described on the label as "beef patties" or "chicken patties" are not considered as having a standard of identity by the USDA. Standardized products, as defined by the USDA, are meat and poultry products for which specific guidelines exist regarding their meat content as well as permitted or prohibited ingredients. Products labeled "hamburger," "hamburger patties," or "chopped beef patties" are considered standardized meat products and as such remain subject to limitations regarding permissible meat ingredients, seasonings, fat, and water content. Added water, phosphates, binders, and textured soy protein products are prohibited for use in hamburger, hamburger patties, and chopped beef formulations. In contrast, beef patties, a nonstandardized meat product, may contain added water, phosphates, binders, and extenders such as various textured vegetable protein products. Other minor differences between the two classes of ground beef products exist; however, for the purpose of the current discussion, beef patties may contain numerous added ingredients, whereas hamburger, ground beef, chopped beef, and both hamburger and chopped beef patties may contain only skeletal muscle meats and seasoning ingredients.

A common question asked when introducing the use of textured soy protein ingredients refers to the quantity of textured soy protein material that can be added to a given meat product. Addressing this issue requires an understanding of both commercial and consumer expectations of the product with regard to final product flavor and texture attributes as well as commercial cost and processing yield considerations. Formulation cost is generally the single most important factor considered when altering meat formulations to include a textured soy protein ingredient. Reducing both formulation cost and cooking shrinkage are important reasons to include textured soy proteins in precooked meat foods. Cost reduction results from replacing the more expensive lean meat ingredients with a less-expensive hydrated textured soy protein extender. Additionally, there is usually sufficient fat in the lean meat being replaced with a textured soy protein material devoid of fat such that some of the lean meat must also be replaced with meat ingredients of higher fat content so as to maintain a similar eating quality as an all-meat product. This provides further formulation cost savings over that of simply replacing meat with hydrated textured soy protein, as fatter meats are valued less than lean meat ingredients. This scenario is valid for products produced and sold in the United States and much of the world; however, in some parts of the world (certain regions in Asia for example), fat meat ingredients cost as much or in some cases more than lean meat ingredients. In such cases, the cost savings would be limited to primarily the cost differential between meat and hydrated textured soy protein. Cost savings associated with replacing meat with a hydrated textured soy protein material are described in Table 9.4.

TABLE 9.4

Beef Patty Formulated To Contain Various Levels of Hydrated Textured Soy Protein Concentrate

Treatment	Raw Ingredient Cost ($/lb; $/kg)	Cooking Yield (%)	Cooked Patty Cost ($/lb; $/kg)	Cooking Duration (minutes)	Formulation Beef 90% Lean Meat Content (%)	Formulation Beef 50% Lean Meat Content (%)
All beef	0.91; 2.00	70.5	1.29; 2.83	6.9	82.2	17.5
80% beef and 20% hydrated textured soy protein concentrate	0.75; 1.65	75.7	0.99; 2.18	6.7	58.0	21.7
70% beef and 30% hydrated textured soy protein concentrate	0.66; 1.45	77.2	0.85; 1.87	6.6	45.8	23.9
60% beef and 40% hydrated textured soy protein concentrate	0.56; 1.23	78.2	0.72; 1.57	6.3	33.6	26.1

Note: Formulation fat content was maintained at 16% across all treatments; textured soy protein concentrate was hydrated with three parts water to one part dry textured soy protein concentrate (weight-by-weight basis).

In addition to product formulation cost savings, textured vegetable proteins improve cooking yields, as shown in Table 9.4. Additionally, it is important to note that the inclusion of textured soy protein products in meat formulations results in increased moisture retention and overall improved eating quality through the initial cooking process as well as through any subsequent freezing and reconstitution.[1] The moisture and fat retention characteristics help to ensure consumer acceptance of many commercial precooked meat products. Formulation costs saving compounded by improved cooking yields, as well as enhanced consumer acceptance, are extremely important benefits associated with the manufacture of precooked meat products such as precooked beef patties and breaded chicken patties served in food service.

From the standpoint of food safety, a growing trend is for meat products to be cooked in a central factory, frozen or refrigerated, and shipped to a restaurant, where they will be heated and assembled into a food product. This processing and distribution concept eliminates or minimizes the chance of raw foods being in close proximity to fully cooked, ready-to-eat foods, thus reducing dramatically the potential transfer of food pathogens from raw to cooked foods. Ingredients such as textured soy proteins that improve the quality of precooked foods will be required to further this food-safety-enhancing initiative.

The beef patty example discussed earlier served to demonstrate the concept of a common meat product in which the permitted product name allowed rather liberal use of added ingredients such as phosphate and textured vegetable proteins with no required minimum meat content other than the requirement that meat must be the most prominent ingredient in the formulation; however, many meat products have a required minimum meat content. The product described by the term "meatballs," for example, may contain no less than 65% raw meat and may contain no more than 12% binder or extender content.[9] Table 9.5 provides examples of minimum meat content requirements for various meat products. It should be noted that, when interpreting USDA meat and poultry regulations, whenever a given minimum meat content is specified for a red meat product this minimum content is formulation raw meat content; however, exceptions can be found.[9,10] The minimum poultry content specified for all poultry products is based on cooked meat content (CFR 9 Part 301.2; CFR 9 Part 381.117(d)). Complicating the formulation of U.S. meat products further, standardized meat products, in addition to almost always having a minimum meat content, also have a limit regarding the quantity of textured vegetable protein that may be included in meat product formulations. In these cases, the USDA provides guidelines for meat substitution for both raw and cooked meat products, and, as described earlier, standards for red meat products vary to some extent compared to those for poultry products. The rule governing the use of textured vegetable protein products in both raw and cooked meat products is known in the meat industry as the *ratio rule*.

TABLE 9.5

Minimum Meat Content Requirements for Selected Meat and Poultry
Products Produced and Sold in the United States

Product	Requirement
Beef burrito	At least 15% raw meat
Beef and gravy	At least 50% cooked beef
Beef tamales	At least 25% raw beef based on total weight of tamale ingredients
Brick chili or condensed chili	At least 80% raw meat
Brick chili with beans	At least 50% raw meat; cereal limited to 16%
Chili con carne	Not less than 40% raw meat
Chili with beans	Not less than 25% raw meat
Poultry burrito	At least 10% cooked poultry meat
Poultry salad	At least 25% cooked poultry; natural proportions of skin permitted
Poultry soup	At least 2% cooked poultry meat
Poultry tamale	At least 6% cooked poultry meat
Poultry turnover	At least 14% cooked poultry meat

The USDA requires controlled or limited use of textured vegetable proteins because these products can simulate the appearance of meat. Unregulated use of textured vegetable proteins could be deceptive to consumers with regard to the meat content of the food; therefore, the USDA maintains strict governance regarding the use of textured vegetable protein products in almost all meat products.

Standardized meat and poultry products containing textured soy proteins (all textured vegetable protein products) must be formulated to strict guidelines regarding textured soy protein content. For cooked meat products, if the ratio of cooked meat to dry (unhydrated) textured vegetable protein material is 9:1 or greater, the product is referred to by its common or usual name.[9] This is the ratio rule. For example, consider a barbeque sauce with chicken product in which the final cooked product contains the minimum required cooked chicken content of 15%; for every nine parts of cooked chicken contained in the product, one part of dry textured vegetable protein may be included in the final product. In this example, the final product could contain only 1.68% dry textured soy protein concentrate, and the product could be labeled "Barbeque Sauce with Chicken." Inclusion of the name of the textured vegetable material (in this example, soy protein concentrate) would be required in the product ingredient statement only. USDA labeling laws permit the use of additional textured protein over that of the 9:1 ratio; however, in such cases, the product name must contain the name of the extender.[9] Under these conditions, should the ratio of cooked meat to textured vegetable protein product be less than 9:1 but not less than 7:1, the product name would be required to contain a qualification statement declaring the name of the textured vegetable protein contained in the product.

Additionally, the letter height of the qualifying statement would be required to be at least one third that of the letter height of the product name.[9] Continuing with the barbeque sauce with chicken example, inclusion of additional textured vegetable protein material such that the final product formulation contained 1 part dry textured vegetable material to 7.5 parts cooked chicken would require declaration of the textured vegetable material in the product name. The product containing the additional textured soy protein concentrate would be labeled "Barbeque Sauce with Chicken with textured soy protein concentrate." The "Barbeque Sauce with Chicken" portion of the name would have a standard letter height, and the qualifying statement "with textured soy protein concentrate" would have a minimum letter height one third that of the first portion of the product name. Also, there are provisions for adding textured vegetable material beyond the ratio of 7 parts cooked meat to 1 part dry textured vegetable protein material.[9] The higher inclusion levels of textured vegetable material requires a declaration of the textured vegetable protein material as part of the product name in such a way that the name identifies significant use of textured protein in the final product. Again, returning to the barbeque sauce with chicken example, inclusion of 1 part dry textured soy protein concentrate for every 6 parts cooked meat would require the name of the product to be "Barbeque Sauce with Chicken and Textured Soy Protein Concentrate Product." In this scenario, the entire phrase on the label would be of the same letter type and letter size. Similar rules apply for textured vegetable protein use in raw meat products. The USDA labeling requirements summarizing the use of textured vegetable proteins in both raw and cooked meat and poultry products are provided in Table 9.6.[9] Note that products described as patties and pizza toppings are exempt from this classification.

The USDA permits some leniency regarding ingredient declaration when both textured soy protein concentrate and powdered soy protein concentrate products are contained in a single meat product. When both textured and powdered soy protein concentrates are contained in meat or poultry products, it is permissible to declare only the name "soy protein concentrate" in the ingredient statement provided both ingredients are permitted in the meat or poultry product. It should be noted that the preferred declaration would list both ingredients (i.e., textured soy protein concentrate and soy protein concentrate) in the product ingredient statement. Also note that the term "powdered" is not used in ingredient statements. Several additional constraints apply, as well. Neither soy concentrate ingredient can be fortified, and the textured soy protein concentrate cannot contain caramel coloring or other colorants. Moreover, no provisions are made for similar single-name declarations for the combination of textured and powdered soy flour products or for the combination of textured and powdered isolated soy protein products. References helpful in understanding USDA meat manufacturing and labeling requirements include the U.S. Code of Federal Regulations (CFR) 9 Parts 317, 318, 381, and 424; the USDA *Food Standards and Labeling Policy Book*; and USDA policy memoranda.

TABLE 9.6

Summary of Use of Textured Vegetable Protein Meat Products Having a Standard of Identity

Type of Meat Used	Meat-to-Dry Soy Ratio	Product Name	Ingredient Statement Declaration
Raw meat	13:1 or greater	Chili con carne	Textured vegetable protein in normal order of predominance
	Lower than 13:1 but at least 10:1	Chili con carne, textured vegetable protein added	Textured vegetable protein in normal order of predominance
	Lower than 10:1	Chili con carne and textured vegetable protein	Textured vegetable protein in normal order of predominance
Cooked meat	Greater than 9:1	Chili con carne	Textured vegetable protein in normal order of predominance
	Lower than 9:1 but at least 7:1	Chili con carne, textured vegetable protein added	Textured vegetable protein in normal order of predominance
	Lower than 7:1	Chili con carne and textured vegetable protein	Textured vegetable protein in normal order of predominance

Raw sausage products made or sold in the United States typically do not contain textured vegetable protein products. Inclusion of textured vegetable protein in raw or cooked sausage would require declaring the names of the textured vegetable protein ingredients in the meat product name. For this reason, textured vegetable protein products are not commonly utilized in fresh, cooked, cured, or smoked sausage products sold in the United States. Again, some exceptions may be granted. For example, if the textured vegetable protein product can be ground sufficiently fine during processing of the meat product, such that the textured vegetable particle is essentially destroyed, the textured vegetable protein ingredient may be considered a binder. Under such conditions, the textured vegetable protein would be listed in the ingredient statement only, without reference to the ingredient having been a texturized material. For example, if textured soy protein concentrate were to be added to a frankfurter formulation and the material was chopped or minced such that there was no evidence of textured soy concentrate being in the cooked frankfurter, then the textured soy protein concentrate ingredient would be labeled as "soy protein concentrate" in the product ingredient statement. With some exceptions, cooked sausage products made and sold in the United States may contain up to 3.5% binder material. Generally, only special circumstances would warrant use of a textured vegetable protein material as a binder in a cooked sausage product and would be subject to USDA approval on a case-by-case basis. Binders are not permitted in fresh sausage product sold in the United States; therefore, no scenarios exist under which textured vegetable proteins would be permitted in fresh sausage products.

People's Republic of China

People's Republic of China food regulations do not provide specific recommendations or rules regarding the use of textured vegetable proteins in meat products. Textured proteins may be used, without much constraint, as a binder in a finely comminuted product or as a particle in coarse ground products; however, specific regulations govern the quantity of total protein (not specifically meat protein) required in specific meat products. The product having the most stringent requirements for ingredient or nutrient content is a product referred to as "retort ham sausage." This product has strict minimum total protein requirements and starch content limitations. Moreover, this product is sold under three grade standards: premium, superior, and common. Retort ham sausage of premium grade must contain at least 12% total protein and no more than 6% starch. Superior-grade sausage must contain at least 11% total protein and no more than 8% starch. Common-grade ham sausage is required to have a total protein content of no less than 10% and must be comprised of no more than 10% starch. Fat content of the retorted sausage is considered self-limiting as it is difficult to include more than 16% fat in the sausage without fatting out during retort processing. Related to the starch and protein requirements, total moisture content is limited to 70, 67, and 64% for premium, superior, and common grades, respectively. (The information provided in this section was derived from a translation of the People's Republic of China individual specification for ham sausage.[10])

Canada

Canadian standardized meat products are required to contain a minimum protein content derived from meat. The minimum meat protein content is described as per a cooked meat weight basis unless specified otherwise. Unlike U.S. meat products, both red meat and poultry products are controlled in a similar manner. According to the 1990 Canadian Meat Inspection Regulations, raw meat products containing fillers for which no required standards exist must contain at least 9.5% meat protein and 11% total protein; cooked meat products must contain at least 11.5% meat protein and 13% total protein. Unlike other countries, Canadian meat products to which a standard of identity applies must include the content of meat-derived protein as part of a product's common name. This tends to limit the addition of fillers such as textured soy protein products. For meat products containing fillers and extenders, such as textured soy protein products, neither the filler nor its content is required as part of the common name of the product, provided fillers are permitted as an ingredient of that product. Meat product filler content is considered self-limiting because all meat products have a minimum meat protein content that must be listed as part of the product name on the principle label panel. Examples of meat products to which

TABLE 9.7

Minimum Meat Requirements for Selected Canadian Standardized Meat Products

Meat Product	Product Minimum Meat Protein Content	Comments
Meat patty	15% (raw meat basis)	No filler/extender permitted
Meat burger	11.5% (raw meat basis) 13.5% (cooked meat basis)	Filler/extender permitted
Cooked meatballs	11.5% (cooked meat basis)	Filler/extender permitted
Sausage (ready-to-eat)	9.5% (cooked meat basis and minimum total protein content of 11%)	Filler/extender permitted
Meat roll	15% (cooked meat basis)	Filler/extender permitted
Toutierre	11.5% (cooked meat basis)	Filler/extender permitted
Meat croquette	11.5% (cooked meat basis)	Filler/extender permitted
Fresh sausage	7.5% (raw meat basis and minimum total protein content of 9% when sold as raw product)	Filler/extender permitted
Breakfast sausage	7.5% (raw meat basis and minimum total protein content of 9% when sold as raw product)	Filler/extender permitted
Potted meat	7.5% (raw meat basis)	—
Creton	11.5% (cooked meat basis)	Filler/extender permitted
Meat pie	Minimum meat content of filling to be 20% (raw meat basis)	Filler/extender permitted
Chili con carne	Minimum 20% of the meat product ingredient calculated as raw meat; additional requirements include a 14% minimum protein content for mechanically separated meat (raw meat basis)	Filler/extender permitted

Canadian Food Inspection Agency standards of identity apply are described in Table 9.7.[11]

Japan

Japanese food ingredients and manufacturing are regulated under the Sanitation Law. Textured soy proteins are commonly incorporated in ground meat products and are listed as soy protein in a raw material statement. Textured plant protein is one of the three forms of plant proteins categorized under Japanese Agricultural Standards (JAS). It is important to note that JAS approval is not mandatory; however, to ensure consumer acceptance, JAS approval is considered important. The JAS mark on food packages provides consumer assurance of both food safety and quality. JAS defines two grades in each of three ground meat products: chilled hamburg, hamburger patty, and chilled meatballs. If the finished products such as patties and meatballs are JAS products, plant proteins in these products must also meet JAS. Japanese product standards are summarized in Table 9.8.

TABLE 9.8

Japanese Agricultural Standards

Product	Grade	Meat Content	Plant Protein (%)	Non-Meat Binders (%)
Chilled hamburg	Premium	≥80 (must be ≥30% beef)	≤10	≤10
	Standard	≥50	≤20	≤15
Hamburger patty	Premium	≥95 (must be 100% beef)	0	0
	Standard	≥75 (must be ≥50% beef)	≤20	≤5
Chilled meatballs	Premium	≥70	≤15	≤15
	Standard	≥50	≤20	≤15

European Union

Uniform guidelines for meat ingredient regulations have been drafted for the European Union (EU); however, at the present time, each country maintains its own meat regulations and meat labeling standards.

Meat Product Formulations Containing Textured Soy Protein Ingredients

Preparation Procedure for Raw Frozen Beef Patties

Table 9.9 provides a U.S. formulation for raw frozen beef patties using encapsulated salt and meat ingredients preground to the final required particle size. When encapsulated salt has been added, it is not advisable to grind the meat mixture in preparation for blending as grinding will crack the protective lipid coating around the salt particle. Exposure of meat pigments to salt during frozen storage will promote pigment oxidation that will appear as undesirable brown or gray blotches against a red or pink background.

Meat and meat blends should be maintained as cold as possible throughout grinding, blending, and packaging. Initial preparation involves grinding the boneless meat through a 1/8-in. (3-mm) grinder plate to prepare the meat for blending and further processing. The textured soy protein concentrate should be prehydrated using three parts water to each part dry textured material; alternatively, hydration can be accomplished during blending, but this is a less desirable method. The ground meat and textured soy protein concentrate should be blended for one to two minutes prior to addition of other ingredients. After mixing only the ground meat and hydrated textured soy protein, the remaining ingredients are added and blended for about a minute. The blending duration depends on the equipment used for blending and the desired meat consistency for forming. Carbon dioxide snow may be

TABLE 9.9

Raw Frozen Beef Patties (U.S. Formulation)

Ingredient	Content (%)
Beef (85% lean, 15% fat)	50.00
Beef (50% lean, 50% fat)	24.40
Water (hydration water for textured soy protein concentrate	18.00
Textured soy protein concentrate (1/8-in. [3-mm] uncolored flake)	6.00
Encapsulated salt	0.80
Alkaline phosphate	0.15
Hydrolyzed vegetable protein (provides flavoring; declaration of plant source required in ingredient statement)	0.45
Ground black pepper	0.10
Onion powder	0.10
Total	100.00

used to bring the targeted mixture temperature to 28 to 32°F (–2 to 0°C) in preparation for forming. Cold blend temperatures ensure that the meat mixture can be formed to the desired shape and size using commercial forming equipment. The final preparation entails forming the meat into the desired shape and product weight followed by individual quick freezing (IQF) and packaging.

The formulation described prepares a beef patty for sale as a raw frozen product. The textured soy protein concentrate in this situation would be uncolored so as to appear as fat in the final raw frozen product. The use of encapsulated salt ensures that no extraction of salt soluble protein from the lean meat will occur during blending and forming. Mixtures containing encapsulated salt should not be ground after blending. As described previously, grinding would crack or break the fat coating the salt granule, allowing oxidation of the meat pigment during frozen storage.

Commercially, textured soy protein concentrates are often hydrated in the blender prior to the addition of meat; however, a common industry practice is to hydrate the textured soy protein concentrate after the coarse ground fat and lean meats have been blended to form a uniform meat mixture by dumping the dry textured soy material on top of the blended meat and subsequently dumping the hydration water on top of the dry soy prior to starting the blender. This reduces total production time, reduces labor, and ensures that water is not lost through blender doors, bearings, etc., as is common in older equipment.

Preparation Procedure for Precooked Beef Patties

Table 9.10 provides a U.S. formulation for precooked beef patties. The formulation and preparation are similar to those for extended meat patties found throughout the world. As described previously, meat and meat

TABLE 9.10

Precooked Beef Patties (U.S. Formulation)

Ingredient	Content (%)
Beef (85% lean, 15% fat)	50.00
Beef (50% lean, 50% fat)	24.80
Water (hydration water for textured soy protein concentrate)	18.00
Textured soy protein concentrate (1/4-in. [6-mm] caramel-colored flake)	6.00
Salt	0.40
Alkaline phosphate (sodium tripolyphospate blend)	0.15
Hydrolyzed vegetable protein (provides flavoring)	0.45
Black pepper	0.10
Onion powder	0.10
Total	100.00

blends should be kept as cold as possible throughout grinding, blending, and packaging. In preparation for blending, the boneless meat should be ground through a 1/4- to 1/2-in. (6- to 12-mm) grinder plate. The textured soy protein concentrate may be prehydrated using three parts water to each part dry textured material (weight-by-weight basis). The coarse ground meat should be blended with caramel-colored textured soy protein concentrate for 1 to 2 minutes, after which the remaining ingredients, with the exception of the salt, can be combined with the blended meats and blended an additional minute. The salt should be added when only 30 seconds of blending remain. The blending duration will be dependent on equipment used for preparation and the desired meat consistency for forming. Blending for longer durations in the presence of salt will make the ground meat texture firmer and less crumbly, thus creating a texture more characteristic of sausage than ground beef. Following blending, the targeted mixture temperature should be between 26 and 28°F (−3°C). Carbon dioxide snow can be utilized to adjust the mixture temperature accordingly. Attaining the 26 to 28°F (−3°C) mixture temperature after blending is necessary because after grinding it is desirable, from the standpoint of forming, to have the meat mixture temperature no higher than 32°F (0°C). Grinding raises meat mixture temperature between 0 and 4°F (0 to 2°C). The meat mixture can be ground through a 3/32-in. to 1/8-in. (2.5-mm to 3-mm) grinder plate and formed into patties using appropriate forming equipment. After forming, patties are cooked using a humidity-controlled cooking chamber, frozen via an IQF process, and packaged. The cooked color would be the brown typical of a cooked beef patty. Use of an uncolored textured protein concentrate would produce a cooked beef color that is lighter brown than that of a beef patty containing only beef or similar beef patty containing caramel-colored textured soy protein concentrate.

TABLE 9.11

Chicken Patties (U.S. Premium Formulation)

Ingredient	Content (%)
Chicken white meat	63.00
Water	20.00
Chicken skin (from white meat trim)	9.25
Textured soy protein concentrate (1/8-in. uncolored flake)	6.00
Salt	0.60
Natural or artificial poultry flavoring	0.50
Alkaline phosphate (sodium tripolyphosphate or commercial blend)	0.35
Monosodium glutamate	0.14
Onion powder	0.06
Garlic powder	0.05
Ground white pepper	0.03
Ground celery seed	0.02
Total	100.00

The formulation described here prepares a beef patty for sale as a pre-cooked and frozen product. The textured soy protein concentrate used in precooked beef patty formulations usually contains caramel coloring and appears as cooked lean in the final product. Beef patty products intended for sale as raw products should contain an uncolored textured soy protein material; the textured protein will appear similar to that of ground fat particles. Note also that in the precooked beef patty formulation, regular ingredient salt replaced the encapsulated salt that was used in the raw beef patty formulation. Encapsulated salt costs over ten times more than the fine flake salt used for most processed meats products. Blending for only a short duration after salt addition favorably minimizes meat protein extraction.

Preparation Procedure for Chicken Patties

Table 9.11 provides a U.S. formulation for chicken patties. Boneless chicken breast meat and chicken skin should be kept as cold as possible throughout grinding, blending, and packaging. Boneless chicken breast meat and chicken skin should be ground through a 1/8-in. (3-mm) grinder plate to prepare the meat for blending. The textured soy protein concentrate should be hydrated using three parts water to each part dry textured material (weight-by-weight basis). The ground meat and skin should be blended with the hydrated textured soy protein concentrate for 1 to 2 minutes. After the ground meat and textured soy protein concentrate are mixed, the remaining ingredients, with the exception of the salt, can be incorporated into the meat mixture. The blending duration may be between 1 and 2 minutes. The salt is added, and the mixture is blended for 30 seconds. The blending duration will depend on the equipment used for blending and the desired meat consistency for forming. Firmer chicken patty textures will be obtained by blending the meat mixture in the presence of salt for longer durations. Carbon dioxide snow can be incorporated during blending to bring the mixture temperature to the targeted temperature of 28 to 32°F (−2 to 0°C) after blending so as to ensure the meat mixture can be formed to the desired shape and size using commercial forming equipment.

Immediately after forming, the patties are battered and breaded and then par-fried in 370 to 380°F (188 to 193°C) frying oil for 30 seconds. The patties are next cooked to a minimum internal temperature of 165°F (74°C) using a humidity-controlled oven. The USDA requires that uncured meats that do not contain nitrite must be cooked to an internal temperature of 165°F (74°C). Industry standard practice is to cook to internal temperatures of as high as 185°F (82°C). Cooked product should be frozen via IQF and packaged. The salt concentration may be altered to provide the desired taste. Additionally, chicken patty texture firmness or hardness may be increased by increasing the salt concentration or by adding the salt ingredient in the early portion of the blending process.

Meat products, sold in the United States that are referred to as "breaded" may contain no more than 30% bread coating; this is calculated as a function of the breaded weight of the product and not as a function of the uncoated meat. Products comprised of greater than 30% batter and breading are labeled as "fritters"; again, breading content is calculated on a final breaded weight basis. An all-dark-meat patty can be prepared substituting the breast meat or light meat trim with thigh meat or dark meat trim. The USDA limits the quantity of chicken skin that can be added to chicken patties to natural proportions relative to the formulation content of the chicken breast meat. The natural proportion of skin covering the chicken breast is considered to be 15%.

Preparation Procedure for Beef Meat Loaf

Table 9.12 provides a U.S. formulation for beef meat loaf. Meat should be maintained as cold as possible throughout grinding, blending, and packaging. Boneless beef should be ground through a 1/4-in. (6-mm) grinder plate to prepare the meat for blending and further processing. The textured soy protein concentrate may be prehydrated using three parts water to each part dry textured material (weight per weight basis). The ground meat should be blended with the phosphate ingredient solubilized in a small amount of water for 30 to 60 seconds followed by the addition of salt and blending for an additional 2 to 3 minutes. In this example, a product with firm texture would be prepared. Blending only the lean meat portion of the formulation with the phosphate and salt prior to the addition of the remaining ingredients should enhance meat protein extraction. If a softer cooked product texture is desired, the ground beef should be blended with the salt for only a short duration. The hydrated textured soy protein concentrate is added to the blended meat followed by an additional one to two minutes of blending.

TABLE 9.12

U.S. Beef Meat Loaf

Ingredient	Content (%)
Beef (50% lean, 50% fat)	37.00
Beef (85% lean, 15% fat)	29.00
Water	18.00
Textured soy protein concentrate (1/4-in. uncolored flake)	6.00
Ground bread crumbs	5.00
Whole egg (frozen or pasteurized)	3.00
Salt	0.60
Hydrolyzed vegetable protein or beef flavoring (optional)	0.50
Dehydrated onion flakes	0.50
Alkaline phosphate (sodium tripolyphosphate)	0.25
Black pepper	0.15
Total	100.00

The remaining ingredients are added to the blended mixture and blended for about a minute. The blending duration will depend on the equipment used for both blending and portioning to achieve the desired product texture. Carbon dioxide snow may be utilized to maintain particle definition during blending, pumping, and portioning. The final meat mixture is ground through a 3/32- to 5/16-in. (2.5- to 8-mm) grinder plate and then transferred to a filling pump so the desired quantity of the meat mixture can be metered into loaf pans. If desired, a tomato paste or puree may be poured or metered over the top of the raw beef meat loaf, which is then baked to an internal temperature of 170°F (77°C).

Preparation Procedure for Beef Chili

Table 9.13 provides a formulation for beef chili. For chili-type products, meat is typically ground through a 3/32- to 1/4-in. (2.5- to 6-mm) grinder plate. Following grinding, the meat should be transferred to a steam-jacketed kettle and cooked using high-temperature steam heat until the desired brownness and particle hardness are achieved. Stirring continuously during cooking will produce very small meat particles; if larger meat pieces are desired, the meat should be cooked with minimal stirring. At this point the cooked meat can be removed from the cooking kettle, leaving the rendered fat behind for sautéing the chopped, fresh onions; alternatively, the fresh onions can be added to the cooked meat. Onions should be cooked until the chopped onion pieces appear translucent. After the onions have been cooked, all remaining ingredients, with the exception of the starch, can be added to the cooking kettle and cooked to hydrate and soften the textured protein and meld flavors. Starch should be added to the cooked meat mixture in the form of a low-viscosity slurry. The mixture should continue to cook as required to

TABLE 9.13

Beef Chili

Ingredient	Content (%)
Beef (80% lean, 20% fat)	45.00
Soaked red beans or drained canned red beans	20.00
Water	12.33
Tomato puree	10.00
Ground or diced fresh onions	5.50
Textured soy protein concentrate (1/8-in. [3-mm] caramel-colored granules)	3.25
Salt	1.30
Whole wheat flour	1.00
Hydrolyzed vegetable protein or beef flavoring (optional)	0.50
Ground chili powder	0.50
Ground cumin	0.20
Garlic powder	0.20
Black pepper	0.15
Ground oregano	0.05
Ground red pepper	0.02
Total	100.00

thicken the chili mixture. Because some moisture loss will occur throughout the entire cooking process, additional water may have to be added to attain the desired consistency. After addition of the starch and possibly additional water, the chili mixture should be brought to a targeted temperature, such as 180°F (82°C). The appearance of large meat pieces can be produced in this product by mixing the finely ground fresh meat with the hydrated soy and salt and cooking with only minimal stirring. This will form large cooked meat chunks in the cooking kettle that can withstand stirring.

Preparation Procedure for Reduced-Fat Pepperoni

Table 9.14 provides two formulations for a product labeled reduced-fat pepperoni. Formulation 1 describes the use of textured soy protein concentrate. Formulation 2 uses an alternate frozen or chilled isolated soy protein gel instead of the textured soy protein concentrate. The isolated soy protein gel is made as described earlier in the text by chopping or blending 1 part isolated soy protein powder with 2.5 parts water (weight-by-weight basis) followed by tempering the mixture until it is quite firm. The firm gel is ground prior to blending into the pepperoni formulation.

Generally, previously frozen meat is used for the manufacture of dried fermented sausages. Such ingredients are utilized for several reasons. Previously frozen meat retains less moisture than meat that was never frozen, thus the product loses moisture more rapidly during drying following fermentation. Moreover, very cold meat that has been tempered to a firm state but not thawed completely maintains good fat and lean particle definition

TABLE 9.14

Reduced-Fat Pepperoni

Ingredient	Formulation 1 (%)	Formulation 2 (%)
Pork (90% lean meat content)	54.30	54.30
Pork (42% lean meat content)	12.50	12.50
Beef (50% lean meat content)	14.00	14.00
Textured soy protein concentrate	5.00	0.00
Textured isolated soy protein granule	0.00	15.00
Water (for textured soy protein concentrate hydration)	10.00	0.00
Salt	3.20	3.20
Paprika	0.24	0.24
Dextrose	0.24	0.24
Cure salt (6.25% nitrite)	0.20	0.20
Red pepper	0.12	0.12
Allspice	0.12	0.12
Anise seed	0.06	0.06
Garlic powder	0.02	0.02
Total	100.00	100.00

Note: Traditional pepperoni contains 30 to 35% fat prior to drying. This reduced-fat pepperoni example contains 17.7% fat prior to drying. Labeling a product "reduced fat" requires as a minimum 25% reduction of fat content.

throughout grinding, blending, and filling into casing. Preparing meat mixtures at warmer temperatures allows the fat particles to smear during blending, grinding, and filling which, in addition to being unattractive, interferes with the required moisture loss during drying. Warmer meat temperatures also promote greater myofibrillar extraction during preparation, which has the undesirable affect of promoting water binding by the meat proteins. Reduced drying efficiency has a major impact on production costs as annual inventory turnover is reduced and greater energy costs must be utilized to produce a given quantity of finished product.

Frozen meat should be tempered to a temperature of between 26 and 28°F (−3 to −2°C). Meat and fat colder than this may shatter during grinding, producing irregularly shaped meat and fat particles as well as undesirable fine meat and fat particles. All meats should be ground through a 1-in. (25-mm) diameter grinder plate in preparation for blending. All ingredients, with the exception of the salt and reconstituted starter culture, are blended for 30 seconds. The reconstituted commercial starter culture is added to the combined mixture and blended for 1 minute; salt is then added and the mixture is blended for 1 minute. The mixture should be ground through a 5/32-in. (4-mm) diameter grinder plate and vacuum filled into fibrous casing of the desired diameter. Filled casings are usually hung in the fermentation chambers and equilibrated at 40 to 50°F (4 to 10°C) for 4 to 6 hours prior to initiation of the fermentation process so as to avoid heat shock to the bacterial culture. Raw sausage should be fermented using the temperature conditions recommended for the bacterial culture. Fermentation will take place between

70 and 90°F (21 and 32°C) and 90 and 110°F (32 and 43°C); selecting the fermentation temperature range is related to the bacterial starter culture incorporated into the meat mixture. Pepperoni is usually allowed to ferment until the meat pH is 4.8 or below. After fermentation, the raw sausages are heated and dried in accordance with validated cooking and drying conditions known to kill *Escherichia coli*. The combined cooking and drying processes should yield a product having a moisture-to-protein ratio of 1.6 to 1 or less. The previous processing description outlines procedures for product made in the United States; outside the United States, processing steps designed to kill pathogenic bacteria may not be considered mandatory.

Meat Analog Food Products

Meat analog products present a challenge regarding their formulation, as the meat products they emulate are extremely complex foods relative to both texture and flavor. Meat has very desirable yet critical flavor and texture attributes that must be duplicated in foods simulating meat products. Generally, meat foods are not one dimensional; rather, they can be thought of as providing a complex sensory interaction involving layers of textures, flavors, and aromas. Therefore, rather involved or complex recipes comprised of textured vegetable pieces, binding agents, fats and oils, and flavorings are required to form even marginally acceptable meat analog products. Ingredients that may be included in a given meat analog formulation must be consistent with the vegetarian population segment for which the meat analog product is intended. An entire book chapter or more could be devoted to the formulation and manufacture of meat analog products. Only a basic introduction to the formulation and preparation of meat analog products is provided here.

As mentioned previously, ingredient selection for meat analog products must be considered for various population segments. Vegetarians can be identified or classified as vegan, lacto–ovo, lacto, or ovo vegetarians. Vegan vegetarians do not consume foods containing any animal products; lacto–ovo vegetarians utilize both eggs and milk in their diets; and lacto and ovo vegetarians limit their inclusion of animal proteins to only milk and only eggs, respectively.

Many different vegetable and grain ingredients can be utilized as the primary ingredients or the bodies of the meat analog products; however, only a limited number of food ingredients work well as the binding material holding the various particulate ingredients together in a cohesive mass such as a vegetarian beef-flavored patty. Egg white remains the most common ingredient used to bind textured vegetable protein materials together to form vegetarian patties as well as vegetarian products that simulate whole-muscle foods such as meat chunks and roasts. Very low concentrations of dried egg white are required to hold blended food ingredients together. Dried egg white is used at between 1 and 2% of a meat analog formulation, while liquid egg white is used at between 10 and 20% of a meat analog. Egg white produces a firm irreversible gel when cooked. Vegetarian foods containing egg products are unacceptable to some vegetarians, so alternative binders are necessary. Specific methylcellulose products are ideal ingredients for use as binding material for meat analog products. Such ingredients not only provide excellent adhesion properties, holding the various textured vegetable proteins together in a bound mass, but also provide lubricity during mastication as characterized by fat. Moreover, methylcellulose gels are reversible, becoming firmer with heating and softer when cooled; these unique gelling properties are the opposite of almost all other gelling materials used in foods. For example, starch gels are reversible gels that can become quite fluid when warmed. Similarly, carrageenan and guar gum gels, though quite firm at refrigeration and room temperatures, become quite soft and less viscous when heated. Foods served cold (room temperature or less) may contain carrageen and or starch as the primary binding material holding other gelled ingredients or textured vegetable protein ingredients together. Products served hot require either egg white or methylcellulose for binding particulate ingredients together. All methylcellulose ingredients are not acceptable for use as a binding agent to meat analog products. The methylcellulose products selected for use in meat analog formulations should be quite viscous at relatively low concentrations.

The various textured soy protein materials discussed earlier in this chapter are excellent particulate materials for inclusion in meat analog formulations. Additionally, textured wheat protein products as well as vital wheat gluten and highly functional isolated soy protein powders are common meat analog ingredients. Wheat gluten and isolated soy protein can be counted on to bind ingredients together; however, products that are required to be firm or hard when consumed warm almost always contain egg white or methylcellulose.

The formulation of meat analog products is quite different from that of meat products. Meat inherently possesses highly desirable textures, flavor, and eating enjoyment. The base or predominate ingredients of meat analog products are usually textured vegetable protein products having only mild flavor (in some cases, flavor that must be masked) and water. Distinct meat-like flavors and the qualities of animal fat must be built into analog products. Flavoring meat analog products is an art. Sufficient flavor must be added to mask soy flavors (e.g., bean, cereal). Additionally, specific flavor top notes must be developed so as to emulate flavors of specific meat products. And, finally, the umami flavor characteristics should be enhanced through the addition of glutamate and nucleotides. Excellent natural and artificial meat-like flavors developed to taste like beef, pork, bacon, chicken, etc. have been formulated by various flavor manufactures. Typical flavoring use levels range between 1 and 5% of the total product formulation, and a mixture of two or more flavors is usually necessary to impart the desired flavor profile. Additionally, salt is an extremely critical ingredient that is required to carry more of the overall flavor impact than required for meat products.

TABLE 9.15

Chicken-Flavored Vegetarian Patty Nugget

Ingredient	Content (%)
Water	66.00
Textured wheat protein (strands, shreds, or fibers of wheat protein)	6.75
Textured soy protein concentrate (1/8-in. [3-mm] diameter flake)	6.50
Textured soy protein concentrate (1/4-in. [6-mm] diameter flake)	5.00
Vegetable oil or shortening (soy, canola, etc.)	5.00
Vital wheat gluten	4.00
Natural or artificial flavoring	4.00
Isolated soy protein	1.00
Methylcellulose (high viscosity)	1.00
Salt	0.40
Ground celery seed	0.15
White pepper	0.15
Garlic powder	0.05
Total	100.00

Meat Analog Product Formulations

Preparation Procedure for Vegetarian Patty Nugget

Table 9.15 provides a formulation for a chicken-flavored vegetarian patty or patty nugget. Place two thirds of the formulation water, textured wheat protein, and methylcellulose in a paddle blender and blend using a paddle rotation speed of 12 to 25 revolutions per minute (rpm) until the textured wheat pieces begin to form defined, almost elastic shreds. Add textured soy protein to the blended mixture and continue blending until the textured soy protein has been partially hydrated; this should take roughly 5 minutes. Add the vital wheat gluten and blend until the powdered gluten begins to form strands. Addition of a portion of the remaining third of the formulation water may be necessary to hydrate the powdered wheat gluten to the desired degree. It is important to save some of the water to be added last. Add the isolated soy protein and blend for several minutes for partial hydration of the isolated soy protein; again, the addition of some of the remaining water may be necessary. Add remaining ingredients with the exception of the remaining water and blend to allow uniform distribution of all ingredients. The mixture may or may not have a uniform firm texture at this point. Add remaining water and blend to allow hydration of the various protein ingredients. If the mixture appears not to have taken up all the oil or the mixture possesses a firm cohesive texture, additional water usually provides sufficient hydration of the ingredients to allow coating of the vegetable oil as well as additional mixture viscosity. Carbon dioxide snow may be required to firm the mixture for forming. Formed nuggets can be cooked in a continuous oven or battered and breaded, par-fried, and oven-baked as described previously.

TABLE 9.16
Vegetarian Chili

Ingredient	Content (%)
Water	37.60
Canned chili beans	22.00
Canned pinto beans	13.00
Textured soy protein concentrate (1/4-in. [6mm] diameter crumble or granule)	10.00
Tomato paste	9.00
Soybean oil	4.00
Beef-type flavor (ingredient contains at least 50% salt)	2.00
Sugar	0.60
Paprika	0.30
Chili powder	0.30
Dried minced onion	0.30
Hot spicy sauce	0.25
Onion powder	0.25
Salt	0.20
Cumin	0.05
Oregano	0.05
Black pepper	0.05
Red pepper	0.05
Total	100.00

Preparation Procedure for Vegetarian Chili

Table 9.16 provides a vegetarian chili formulation. Combine water, textured soy protein concentrate, and minced dried onions in a jacketed steam kettle and slowly stir while heating for the purpose of hydrating the textured protein material. Add remaining ingredients and cook, stirring continuously, to meld the flavor contributions of the various ingredients. Cook ingredient mixture until the desired viscosity has been achieved. It may be necessary to add water before the completion of cooking. Generally, between 8 and 15% of the water is lost due to evaporation during cooking of kettle-cooked foods. Package, chill, and freeze chili food material.

Summary

Textured soy protein materials are extremely versatile food ingredients due to their unique meat-like textures after hydration, bland flavor profiles, and amino acid composition that provides a quality of protein similar to that of meat, eggs, and milks. These attributes are responsible for their expanding acceptance as major ingredients in meat and meat analog products throughout the world. This chapter was intended as an introduction with regard to the use of textured soy proteins in foods. It should be noted that textured vegetable protein products derived from wheat as well as other sources are excellent ingredients for consideration for incorporation in meat and meat analog products; these ingredients have interesting properties that can be utilized as is or altered to provide specific textural properties to foods. However, globally textured soy protein products remain the most widely used texturized vegetable protein materials utilized in foods. Manufacturers of textured soy proteins strive to bring new product innovations to the food industry; novel soy protein textured products as well as textured ingredient blends allow improvement of existing meat and meat analog products as well as the introduction of interesting foods that could not be produced otherwise.

Assistance regarding product formulation and textured soy protein nutritional information are readily available through the websites of the various companies manufacturing the various textured soy protein products. This information is also available through the technical and sales departments of the various soy protein product manufacturers. At the time of publication of this book, the major producers of textured soy protein products include The Solae Company, St. Louis, MO; Archer Daniels Midland, Decatur, IL; Cargill, Minneapolis, MN; Solbar Industries, Ashdod, Israel; and Legacy Foods LLC, Hutchinson KS.

References

1. Hargarten, P.G., Hall, P.W., Tolbert, M.A., and Campano, S.G., Use of Textured and Functional Soy Protein Concentrates for Texture Improvement in Chicken Patties, paper presented at Food Ingredients, South America, 1996.
2. Bhattachorya, M., Hanna, M.A., and Kaufman, R.E., Textural properties of extruded plant protein blends, *J. Food Sci.*, 51, 988, 1986.
3. Harper, J.M., *Extrusion of Foods*, CRC Press, Boca Raton, FL, 1981, p. 96.
4. ADM, TVP® Technical Sheet, Archer Daniels Midland, Decatur, IL.
5. Legacy Foods, Product Technical Sheet, Legacy Foods LLC, Hutchinson, KS (http://www.pmsfoods.com).
6. McMindes, M.K. and Richert, S.H., Process for the Production of an Improved Protein Granule Suitable for Use as a Meat Extender, U.S. Patent 5433969, 1995.
7. Parks, L.L. and Greatting, A.D., Process for the Production of a Protein granule suitable for use as a meat extender, U.S. Patent 5160758, 1992.
8. Nguyen, T.V. and Wagner, T.J., Process for the production of a granulated protein gel suitable for use as a meat extender, U.S. Patent 4276319, 1981.
9. USDA, *Food Standards and Labeling Policy Book*, Office of Policy and Program Development, Food Safety and Inspection Service, U.S. Department of Agriculture, Washington, D.C., 2003, p. 4.
10. *Industrial Specifications for Ham Sausage*, People's Republic of China, SB 102 51-2000.
11. CFIA, Meat and poultry products, in *Guide to Food Labeling and Advertising*, Canadian Food Inspection Agency, Ottawa, Canada, 2003, chap. 14 (http://www.inspection.gc.ca).

10

Food Bars

Steven A. Taillie

CONTENTS

Product Market and Positioning

The food bar category has the best average growth rank of all food and beverage categories, with growth estimated at greater than 40%. Although this is primarily a North American phenomenon, the rest of the world may

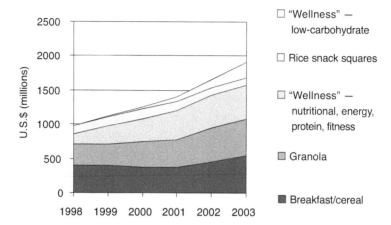

FIGURE 10.1
Evolution in bar categories with time. (Information Resources, Inc., supermarket or food/drugstore/mass merchandiser combined channel review databases, excluding Wal-Mart.[5])

be catching up, although it has a long way to go. Bars were initially positioned in the United States as convenient, portable breakfasts. Later, the wellness, energy, and fitness category took off (Figure 10.1 and Figure 10.2). These products became more interesting to the major confectionery companies as sales in the category grew. Some companies entered the field as a result of losing market share in more traditional lines, driving growth even faster. Most recently, the low-carbohydrate trend has propelled the food bar market to its current level of around $1.9 billion, although the phenomenon seems to have peaked of late. Today, some bars are positioned as a healthy

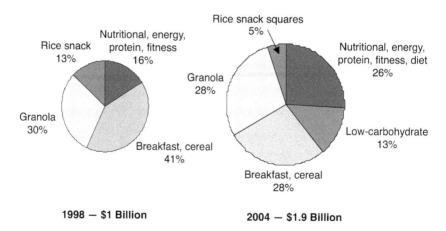

FIGURE 10.2
Evolution of bar market size and composition in the United States from 1998 to 2004. (Information Resources, Inc., supermarket or food/drugstore/mass merchandiser combined channel review databases, excluding Wal-Mart.[5])

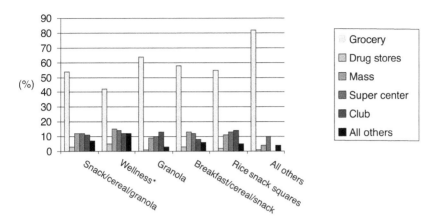

FIGURE 10.3
Bar outlets dominated by grocery stores in 2004. Wellness bars include low-carbohydrate plus nutritional, energy, protein, fitness, and diet bars. (Information Resources, Inc., supermarket or food/drugstore/mass merchandiser combined channel review databases, excluding Wal-Mart.[5])

means of obtaining nutrients for energy or for mental acuity. Developments continue, and emerging products are being positioned as meal replacement products providing particular health benefits. Since the major confectionery manufacturers entered the market, the pattern of sales through the various outlets has shifted. In 1997, natural food stores accounted for 54% of bar sales in the United States; today, grocery stores predominate (Figure 10.3).

Definition

Food bars are combinations of ingredients that provide food in a solid, low-moisture form. They are consumed as a source of nutrients, as opposed to confectionery bars, which are consumed as sweet products. They are formulated to contain a wide range of nutrients and are positioned in various ways for marketing purposes.

Formulation Types

The basic formulations of food bars include:

- Athletic bars
- Lifestyle or wellness bars (includes the 40:30:30 concept of balanced calorie intake from carbohydrates, protein, and fat)

TABLE 10.1

Types of Bar Formulations and Typical Compositions

Bar Formulation	Protein (g/serving)	Carbohydrates (g/serving)	Fat (g/serving)
Athletic	25–40	10–30	0–5
Wellness	12–16	20–25	6–8
Meal replacement	6–8	35–38	0–5
Energy	5–15	25–50	2–6
High-protein	>35	2–6	3–15

Note: In the U.S. market, serving size varies from 28 to 80 g; a typical range is 50 to 56 g.

Source: Data from Solae LLC, St. Louis, MO.

- Diet bars, including meal replacement bars and high-protein, low-carbohydrate bars
- Carbohydrate energy endurance bars, typically used by backpackers, climbers, cyclists, etc. for an energy boost

All food bars use similar combinations of ingredients, but their positioning varies widely in each of the categories listed above. Although the ingredients used may be similar, the nutrient profile in each category is different (see Table 10.1).

Physical Forms

Food bars can be placed in one of four major physical form categories:

- *Granola bar* — This type of bar represents the very first category to appear on the market in the United States. It consists of a combination of whole grains (typically rolled oats and other cereals) bound together in a syrup matrix. The products are produced on a sheeting line.
- *Baked bars* — This type of bar could be a baked granola bar or a product similar to the Kellogg's Nutri-Grain® line. Again, a sheeting line would be used in the manufacture of these bars, but it would be followed by a baking step in a continuous band oven. Baked bars may contain fruit purées or fruit pieces.
- *Nutritional bars* — The nutritional bar market contains two prevalent types: sheet and cut bars (e.g., Clif's Luna™ bar) containing a combination of whole grains, crisp rice, and 50 to 80% protein nuggets, which are a key ingredient. The extruded nuggets are produced by

a twin-screw, cooking extruder. The process expands the soy proteins and results in a texture similar to that of puffed rice. Again, a sheeting process is involved in the manufacture, usually a three-roller system to form the sheet. Conveyers carry the sheet to a rotating disc "pizza-cutter" slicer and then to a guillotine that cuts the slices into the correct length for the bars. Some may then pass through an enrobing step, although usually only bottom enrobing is done.

- *Extruded bars* — For extruded bars, a forming extruder is used, not a cooking extruder as referred to above. The dough matrix is forced through a die of the appropriate dimensions to form ribbons of dough that can then be guillotined into the correct lengths.

Processing

All four of the processing methods for the forms described above start with a mixer. Large-scale operations would call for 500-kg capacities, although very large plants may use 750-kg mixers. These mixers are usually of the double sigma blade type, because high shearing forces are required to cope with the very high viscosity doughs produced. For sheet and cut bars, a gentler type of mixer is used: either a ribbon or a paddle mixer. The high shear action of sigma blade mixers tends to break the crisp rice or soy protein nuggets in the formula.

The ideal factory layout positions these mixers on a mezzanine floor above the rest of the bar line so gravity guides the discharge into the sheeting roll (or extruder) hopper below. Alternatively, if the mixers are on the same level as the rest of the production line, wheeled dough troughs can be used. These convey the doughs to the feed hoppers, into which they are fed manually. Line-belt widths of 1 m are typical of large commercial operations. This size handles 20 to 25 ribbons of dough simultaneously. One or two cooling tunnels may be necessary. The first cooling tunnel sets the bar matrix for ease of handling and is typically found in sheet and cut bar lines. The second tunnel is used to set any compound coating that may have been applied by the enrober. Compound coatings are the most common form used, because they have the advantage over pure chocolate of not requiring tempering to obtain a good set and they avoid later fat blooming problems. Typical compound coatings used are sugar-based milk chocolate, yogurt, or white chocolate. Less commonly used is dark chocolate. For the popular high-protein, low-carbohydrate bars, sugar-free coatings are used. These are formulated with sugar alcohols, such as maltitol, isomalt, or erythritol.

A sheeting line could be equipped with a one-, two-, or three-roll extruder to produce the dough sheet. A second set of sheeting rolls can be used to place a second layer on top of the first one. A protein matrix sheet could be

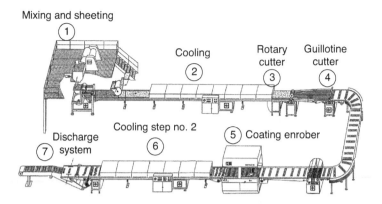

FIGURE 10.4
Typical plant layout for a sheet and cut bar line.

used as a base layer that is covered with nougat or caramel, for example. This procedure is common for confectionery bars. Particulate ingredients can then be sprinkled on top of the second layer (e.g., peanuts or other nut meat pieces) and then rolled into the surface layer (Figure 10.4).

Ingredients

Ingredients can be conveniently classed as follows.

Carbohydrates and Polyols

Carbohydrates and polyols are primarily used to provide the matrix that binds the dry ingredients together. The common forms used are corn syrups or high-fructose corn syrups (HFCS). Of the corn syrups, 42 or 63 dextrose equivalent (DE) products are often used. Some bars can be found with a high-maltose corn syrup. For the HFCS types, typically 42 or 55% fructose products are used. Brown rice syrups and maltodextrins are also used. High-protein, low-carbohydrate formulations obviously require alternatives to these products. In this case, sugar alcohols, or polyols, can be used. Sorbitol, maltitol, lactitol, and erythritol are commonly found. Glycerin and polydextrose are also used, and polydextrose may be labeled as a source of dietary fiber. Hydrogenated starch hydrolysates (HSHs) are also utilized. The use of polyols in formulations requires the inclusion of some form of high-intensity sweetener, as all of the polyols have significantly lower sweetness levels than sugars. Cyclamates, acesulfame-K, aspartame, saccharin, sucralose, and, most recently, neotame are found in bar formulas, and alitame could also be available in the near future. Among these, acesulfame-K and sucralose are probably the most commonly used.

Protein Sources

Soy protein is one of the predominant players in nutritional food bars. It exists in many forms, from 50% protein soy flours to 70% protein soy concentrates to 90% protein isolated soy proteins. In addition to these forms, soy can be added in the form of soy nuggets ranging in protein content from 50 to 80%. Solae LLC (St. Louis, MO) supplies all of the required higher level soy protein forms under a variety of brand names, and an active research and development program supports the regular introduction of new products. Dairy proteins are also commonly used in food bars. Products processed from whey include whey protein concentrates (WPCs) and whey protein isolates (WPIs). Caseinates are also present in some formulas, as are whole milk protein extracts. Whey, nonfat dry milk, and whey protein hydrolysates are also used. Other sources of protein include gelatins, whole and hydrolyzed. Of the non-soy proteins, probably the ones that predominate are the WPCs and WPIs.

Fats and Oils

Hydrogenated or partially hydrogenated shortenings are typically used in food bars. In some cases, soybean or sunflower oil is used. Soy lecithins are also employed, serving the dual role of fatty acid source and emulsifier, acting to prevent the rest of the fat in the bar from oiling out.

Dietary Fiber

Dietary fiber ingredients are sometimes included to allow the addition of a fiber claim to the label. Fructooligosaccharides and other fructose polymers, such as inulin, are utilized. Polydextrose, as already mentioned, can also be labeled as a fiber source, even when used for functional reasons.

Minor Ingredients

Included in this category are vitamins and minerals, added for nutritional reasons in the form of a premix provided by various suppliers and calculated to provide the correct level of supplementation for the label declaration. These levels can range from 20% of the recommended daily allowance (RDA) up to 100% RDA for a variety of nutrients. Food bars also list flavors (natural or artificial) and colors. Among the many flavoring ingredients that can be chosen, cocoa powders, peanut flour, and peanut butter supply unique, characterizing flavors. Natural pigments are often used as colorants, such as beta-carotene, anatto, and anthocyanins. Also found are botanicals, which are extracts of plants that have demonstrated, or supposed, physiological effects on the consumer's sense of well-being.

Impact of Ingredients and Other Parameters on Bar Texture

Isolated Soy Protein

By far the largest proportion of the soy protein used in bars comes in the form of isolates. Isolates contain a minimum of 90% protein on a dry matter basis, by definition. Two main functional types are used in food bars. The first type has a high gel strength. Solutions of this type of protein have high viscosities and form rigid, thermostable gels at moderate concentrations. The second type provides products whose solutions have much lower viscosities and will not form gels at any concentration. In general terms, the highly gelling isolated soy proteins tend to make firmer bars with a drying mouth-feel. The texture tends to be shorter, in the direction of a cookie. The low-viscosity soy proteins tend to produce much softer, chewier nutritional bars, more in the direction of a Tootsie Roll® type of texture. Water-holding capacity (grams water bound per gram protein), viscosity, and wettability all play a role in determining the texture of the end product (Figure 10.5).

Carbohydrates

Corn Syrups

The dextrose equivalent of the syrup used can have a major impact on bar texture. As the dextrose equivalent increases — for example, from 42 to 63 to 95 DE — the bars become softer. This is also true for rice syrup solids and maltodextrins. If HFCS is being used, the percent isomerization of the dextrose to fructose will play a role in determining bar hardness. A 90% fructose syrup will make an extremely hard bar, whereas a 42% syrup will produce a much softer bar. Syrups of 55% conversion are widely available and deliver roughly the same sweetness level as sucrose. Often, blends of corn syrups and HFCS are used. A 55:45 ratio gives consistently good results for a 63 DE corn syrup. As these ratios change, the texture is again affected. Manufacturers typically boil their corn syrups to a specific brix. A standard 80 to 82° brix corn syrup provides consistent results. As the brix increases and more water is lost, the level of soluble solids goes up and bar texture becomes much harder. In water-starved systems, where sugar alcohols are employed, increasing the free water in the formula softens the bars; however, caution must be exercised when free water is added to a bar matrix. Too much free water will increase the water activity (A_w) value to the point where microbial spoilage can occur. The typical A_w range for food bars is 0.45 to 0.65.

Sugar Alcohols

Sugar alcohols have a reduced caloric density when compared to their sugar homologs and are thus suitable in high-protein, low-carbohydrate formulas. Sugar alcohols also have important humectant properties and will act to effectively lower the A_w, thus helping to reduce the risk of mold spoilage.

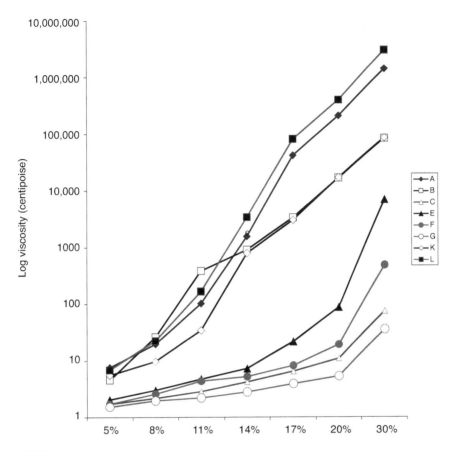

FIGURE 10.5
Different soy proteins produce solutions of different viscosities.

They can be conveniently classified according to their degree of polymerization, like the carbohydrates (Table 10.2 and Table 10.3). Sugar alcohols are poorly digested, resulting in their relatively low calorie values compared to the corresponding carbohydrates. They arrive intact in the large intestine, where bacterial fermentation produces gas. Unabsorbed, intact polyols can also produce osmotic diarrhea. For a normal adult, and depending on the polyol consumed, the tolerated dose is of the order of 0.3 to 0.5 g/kg body weight per day.[1] Experiments using blends of glycerin, polydextrose, maltitol, and sorbitol have indicated that combinations of these polyols can have a profound impact on bar hardness. Control of end product texture can be obtained by judicious blending of these humectants.

pH

A typical bar made with either a hydrolyzed or a nonhydrolyzed isolated soy protein as the protein source generally has a neutral pH. The usual range

TABLE 10.2

Classification of Polyols[2]

Class	Examples
Mono-polyols	Sorbitol
	Mannitol
	Xylitol
	Erythritol
Dipolyols	Maltitol
	Lactitol
	Isomalt
Mixtures	Hydrogenated starch hydrolysate
	Maltitol syrups

TABLE 10.3

Polyol Properties

Polyol	Relative Sweetness[a]	Energy Value (kcal/g)
Xylitol	100	2.4
Maltitol	90	2.1
Maltitol syrups	70	3.0
Erythritol	60–80	0.2
Sorbitol	50–70	2.6
Mannitol	50–70	1.6
Isomalt	45–65	2.0
Lactitol	30–40	2.0
Hydrogenated starch hydrolysate	25–50	3.0

[a] Compared to sucrose = 100.
Source: Adapted from ADA, *J. Am. Diet. Assoc.*, 104, 255, 2004.

would be from 6.7 to 7.0 (5% slurry in water). Food bases, such as sodium bicarbonate, can be used to raise the pH of the bar to 7.3 or 7.4. This has the effect of softening the bar over its shelf-life. Equally, the addition of food-grade acids to the formula (e.g., citric or malic acids) to reduce the pH to around 6.0 to 6.1 has a more dramatic effect, softening the bar texture to an even greater extent (Figure 10.6). The impact of pH on flavor also must be considered. The addition of food bases to chocolate-flavored bars will have a positive flavor impact. The opposite is true for fruit-flavored or fruit-containing bars, which will benefit from the addition of food acids.

Extruded Soy Nuggets as Ingredients

Nuggets are commonly used in bar manufacture as a means of opening up the texture of otherwise dense nutrient masses. They provide a crisp, crunchy bite. Nuggets consist of soy protein extruded through a high-pressure, high-

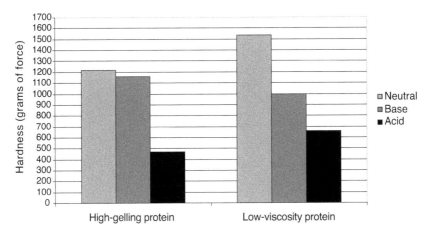

FIGURE 10.6
Impact of pH on bar texture.

temperature cooker extruder to yield a puffed pellet. At the same time, their contribution of protein to the nutrient profile can be important. The extrusion process completely denatures the native soy proteins, reducing their water-binding capacity compared to the more functional isolated soy protein powders, a common part of the matrix. Nuggets are available in protein concentrations ranging from 50 to 80%. Particle density and nugget size can vary, and nuggets are also available that contain fiber and zero net carbohydrates, making them very versatile bar ingredients (Figure 10.6). Nuggets have a limited impact on bar hardening because of the nature of the soy protein contained in them. Nuggets give a prolonged, nonhardened shelf-life when compared to bars made with protein powders. In sheet and cut bars, nuggets are the primary ingredient characterizing texture. In extruded bars, they also provide texture and can be used at levels up to 10%.

Dairy Proteins

Dairy proteins, like their soy counterparts, vary in how they perform in food bars, depending on how they are processed. In general, whey protein isolates make harder bars than whey protein concentrates. Caseinates, both calcium and sodium, or blends of the two can make either harder or softer bars than the whey protein isolates or concentrates, depending on the process and selection of other ingredients. It should not be assumed that all milk proteins in a particular category will perform equally. Care should be exercised when selecting a milk protein product from among the many on the market to find the best one for the application (Figure 10.7). Different types of caseinate and various blends also have a major impact on bar hardening over time. Calcium and sodium caseinates and blends of various ratios have been studied. Processing modifications for each blend selected are another consideration.

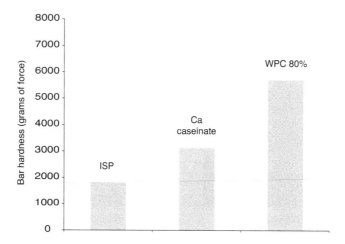

FIGURE 10.7
Impact of milk proteins on bar hardness.

Processing

Various aspects of the production process can have an impact on bar texture. Mix time and speed in the mixer (rpm) seem to be important, as they contribute to the input of shearing forces into the dough. Processing temperature also plays a role. High-viscosity soy protein isolates appear to have a good tolerance to more severe mixing conditions. Low-viscosity proteins do not tolerate such changes in mixing conditions, and texture changes will result (e.g., the bars will tend to harden if overmixed). The order of addition of ingredients to the mixer and so-called "floor time" (in other words, the time the mass spends resting between mixing and the next processing step) can both affect the hardening characteristics.

Shelf-Life Considerations

Manufacturers normally require a shelf-life of 9 to 12 months. Usually, the texture of nutritional food bars changes over time. Textural change generally manifests itself as hardening of the bar. This does not seem to be due to moisture loss, because the same change is observed in food bars that are hermetically sealed in high-barrier films. A shift occurs in the distribution of moisture to move toward equilibrium.[3] This shift changes the product texture. The water absorption properties of the proteins used is a factor of key importance in determining at which textural point the equilibrium position lies. Sensory testing of nutrition bars by expert panels using quantitative descriptive analysis (QDA) techniques[4] and consumers has revealed that the biggest defect in the organoleptic properties of bars is texture change in the direction of hardness over time. Positive features were identified as low cohesiveness, moistness of the bar mass, and moisture absorption. Negative features were described as hardness, roughness of the mass, excessive cohesiveness, and excessive density; however, of all the negative features, hardness was by far the most significant.

Bar hardness can be measured using mechanical force determinations. Instruments employed include texture analyzers of various types, such as Instron's texture analyzers or the newer TAX2 texture analyzer systems. These systems are based on the resistance to the penetration of a rounded puncture probe. The degree of penetration is standardized for a bar of given dimensions. Mechanical hardness can be correlated with sensory data. Many commercial high-protein bars have sell-by dates as much as 12 months or more after production; however, from a practical point of view, panelists can detect changes in texture after 2 to 3 months and will often judge bars unacceptable after only 4 to 6 months on the shelf.

As discussed earlier, the most important factor affecting bar hardness is formulation — the selection of ingredients; however, the level of ingredients also has an impact. The more protein packed into a bar of a given size, the harder the bar will be. Not only is this true initially but also the rate of hardening over time will be faster. The judicious choice of soy proteins can lead to a fivefold reduction in the rate of bar hardening. The effect is particularly evident in high-protein formulations where the protein level is greater than 30%. Dietary fiber ingredients tend to accelerate bar hardening in general because of their high water absorbencies; however, by using the correct blend of proteins, up to 6 or 8% fiber can be included without giving excessive hardening.

Reduction in total moisture content clearly is also a factor in determining initial hardness and the rate of hardening. Also, the types and mixtures of syrups, the pH, and the types and blends of polyols have an effect on texture, as do processing variables such as temperature of mixing, shear, and mix time. The size of the mixer in relation to batch size is also a processing parameter to be controlled, as well as pressure at the extruder step.

Conclusions

As is the case in most complex food systems, the mysteries of controlling the texture of food bars can be understood only by a return to basic scientific principles. The disciplines involved include an understanding of the physics controlling the mass transfer of moisture in complex, multicomponent systems. The mathematical modeling of these systems has been developed only comparatively recently, in the past 20 years or so, and the analytical methods required to peek inside these systems have only recently become available for scientists to test the models. Sophisticated techniques such as magnetic resonance imaging are proving invaluable to our developing an understanding of how these systems behave.

References

1. Storey, D.M., Koutsou, G.A., Lee, A., Zumbe, A., Olivier, P., Le Bot, Y., and Flourie, B., Tolerance and breath hydrogen excretion following ingestion of maltitol incorporated at two levels into milk chocolate consumed by healthy young adults with and without fasting, *J. Nutr.*, 128, 587, 1998.
2. ADA, Position of the American Dietetic Association: use of nutritive and non-nutritive sweeteners, *J. Am. Diet. Assoc.*, 104, 255, 2004.
3. Le Meste, M., Mobility of small molecules in low and intermediate moisture foods, in *Food Preservation by Moisture Control*, Barbosa-Canavas, G.V. and Welti-Chanes, J., Eds., Technomic, Lancaster PA, 1995, p. 209.
4. Hootman R.C., *Manual on Descriptive Analysis Testing for Sensory Evaluation*, American Society for the Testing of Materials, West Conshohocken, PA, 1992.
5. IRI, *Times and Trends: A Snapshot of Trends Shaping the CPG Industry*, Information Resources, Inc., Chicago, IL, 2004.

11

Ready-To-Drink Soy Protein Nutritional Beverages

Paul V. Paulsen, David Welsby, and Xiaolin L. Huang

CONTENTS

Introduction

Markets for Soy Protein Nutritional Beverages

Beverages containing soy protein offer many market opportunities. These can range from high-volume, lower margin products for general consumers (e.g., soymilk or fortified juices) to lower volume, high-margin products (e.g., infant formula or disease-specific medical supplements). Food marketers and beverage manufacturers have found soy to be an economical, healthful, and functionally versatile source of protein. Soy consumption in beverages has a long history in various parts of Asia, where different forms of extract from whole soybeans have been sold as fresh soymilk in local markets for many years. By 1923, a soymilk factory was operating in Changs, China, selling soymilk in bottles.[27,28] The history of soymilk marketing in other world regions is much shorter — 20 years or less. For example, the current market in North America began with the introduction of retorted packs of soymilk from Japan 20 years ago. Although a small market based on sales of products to Asian–Americans and Seventh-Day Adventists existed, the convenience and quality of the imported products soon changed market dynamics. Today, soymilk sales revenues in the United States are over $600 million and soymilk products can be found in a wide spectrum of retail outlets.[39]

Feeding soy to infants and children is also an idea transferred from Asia. Since about 1929, soy-based infant formula (SBIF) has been consumed by infants. Today's generation of infant formula originated in the 1950s with the recognition that infants with milk allergies could be nursed on a beverage formulated with soy proteins. Besides its effectiveness as treatment for IgE-mediated cow milk allergy, SBIF is consumed for lactose intolerance and galactosemia and as a vegetarian human milk substitute.[2] The world soy infant formula industry has multiple pharmaceutical marketers and revenues over $500 million.[8]

Today's consumers increasingly recognize the importance of good nutrition to counter the physical and mental demands of modern life. Issues such as heart health, arterial conditions, adult-onset diabetes, weight management, and cancer have caused consumers to seek foods containing nutritional substances that may ameliorate or prevent their symptoms. Substances such as antioxidants, calcium, prebiotics and probiotics, fiber, and soy proteins have particular relevance to these consumers. The worldwide functional foods market has an estimated value of $1.3 billion.[10] These consumer interests are driving product innovation with soy protein as an important component. Soy proteins are one of the top choices for food product developers interested in bioactives.[33] Soy versions of milk shakes, protein drinks, fruit drinks, and yogurt smoothies are growing in popularity, with products appearing in markets across the world.

Nutritional and Functional Demands

Successfully marketed beverages deliver great taste, convenience, and stability at a reasonable price. In addition, as already mentioned, nutrition has become a differentiating factor among consumers. It can be a difficult task for the food processor to balance these consumer needs. Providing a long shelf-life may address the stability issue but compromise the taste. The addition of calcium may improve nutrition but compromise stability. The beverage processor must weigh the ingredients, processes, and technologies available against established nutritional and functional demands to select a best solution.

The functional demands of a beverage can be categorized into three main areas: sensory attributes, safety attributes, and physical attributes. The sensory attributes can be subcategorized as aroma, flavor, appearance, and mouthfeel. The safety attributes can be subcategorized as microbial toxins or pathogens, manufacturing contaminants, allergen factors, and recognized safe formulations (ingredient use or level); it should be apparent that a universal demand exists for controlling these safety subcategories to zero levels of noncompliance. The physical attributes can be subcategorized as suspension, viscosity, translucency/opacity, and particle size. Beverages are inherently perishable food products, so the beverage manufacturer must ensure that each attribute remains stable with time.

The first element of nutritional demand is the chemical and proximate profile desired for the beverage. A beverage can be grossly defined by its protein, fat, carbohydrate, and fiber composition. These will generally account for 95 to 99% of the solids of a beverage. Also included are smaller nutrient components such as vitamins, minerals, and other bioactive materials. The targeted overall nutritional profile must be considered along with

TABLE 11.1

Human Requirements for Essential Amino Acids (g/100 g protein)

Amino Acid	Infant Mean (Range)[a]	2 to 5 Years	10 to 12 Years	Adult
Histidine	26 (18–36)	19	19	16
Isoleucine	46 (41–53)	28	28	13
Leucine	93 (83–107)	66	44	19
Lysine	66 (53–76)	58	44	16
Methionine and cystine	42 (29–60)	25	22	17
Phenylalanine and tyrosine	72 (68–118)	63	22	19
Threonine	43 (40–45)	34	28	9
Tryptophan	17 (16–17)	11	9	5
Valine	55 (44–77)	35	25	13

[a] Data from FAO/WHO Joint Expert Consultation.[9]

the expected shelf-life and delivery system for the beverage. In most cases, the formulated nutrient components will have to be at higher levels than claimed on packaging to withstand the process and delivery system.

Protein plays a vital role as a nutritional and functional component in many existing beverage products.[44] Many consumers see protein as being critical to muscle development and replacement, but the emphasis may change depending on the specific consumer. For example, the sports nutrition area is becoming more segmented, and products for power sports, such as weightlifting, are different from those for endurance athletes. Weightlifters require protein to build muscle mass, while long-distance runners or bikers want quick replenishment of worn muscles during and after competitions. For people undergoing weight-loss programs, a source of high-quality protein is even more important due to the risk of losing lean body mass at low calorie intakes.

The daily protein allowance is based on the varying amounts of amino acids needed, depending on age, body size, sex, and physiological condition. Extensive research has determined the amino acid requirements for various ages and conditions (see Table 11.1). When the requirements for amino acids are met, the recommended daily protein intakes can be calculated for various ages, genders, and conditions. The U.S. recommended daily allowance (RDA) recommends a range of 2.2 g protein per kg body weight for a young infant to 0.8 g/kg body weight for adults.[11]

Soybean proteins exhibit one of the best amino acid patterns among vegetable protein sources. Raw soybean protein is not easily used due to inhibitory factors such as trypsin inhibitors, urease, and hemagglutinin compounds; however, destruction and removal of these inhibiting factors, as done for commercial sources of concentrated and isolated soy proteins, enables high protein utilization. Protein isolates and concentrates have a higher digestibility than soybean flour.[3] Some isolated soy proteins, when correctly processed, are comparable to other high-quality proteins found in meat, milk, and eggs.[29]

Food proteins vary widely in their functional properties. The beverage developer should exercise care in selecting an appropriate protein for ready-to-drink products. In multiphase systems such as most liquid drinks, the proteins must contribute to flavor, color, mouthfeel, viscosity, mineral suspension, and emulsion stability. Also, the chemical and microbiological control of protein products must be sufficient to prevent the occurrence of safety issues.

Soy Protein Health Benefits

Since 1999, several countries (United States, United Kingdom, South Korea, and Brazil) have approved the use of soy health claims. The standard claim is that "25 grams of soy protein a day as part of a low-fat and low-cholesterol diet may reduce blood cholesterol and the risk of coronary heart disease." Foods that deliver at least 6.25 g soy protein per serving are allowed to make this heart-health claim on their packaging in the United States. Other countries allow package claims regarding cholesterol reduction as a result of soy protein consumption.

Numerous studies have shown a strong association between consuming soy protein and improvement in risk factors associated with cardiovascular disease. A metaanalysis of studies investigating soy protein and cholesterol showed that consuming soy protein has a significant effect on lowering blood cholesterol, especially low-density lipoprotein (LDL) cholesterol. Soy protein intake maintains levels of the more desirable high-density lipoprotein (HDL) cholesterol, which is often reduced on low-fat diets. This is an important finding because one of the strongest predictors of heart disease is a low ratio of HDL to LDL cholesterol.[1] The bioactive compounds that cause this effect have not been categorically identified; however, globulin proteins, peptides, amino acids, phytic acid, saponins, isoflavones, and protease inhibitors have been implicated. A number of studies recently found that isoflavones contribute to maintaining the elasticity of arteries and lessening hardening of the arteries.[31,40] Clinical research continues on the impact of soy protein consumption on heart and vascular health, as well as on cancer and women's health (e.g., osteoporosis and menopausal symptoms).

In 2004, The Solae Company petitioned the U.S. Food and Drug Administration (FDA) to seek approval for a qualified health claim that suggests that the consumption of soy protein may reduce the risk of certain types of cancers. The petition for the health claim focuses on 58 studies supporting an association between soy consumption and reduced risk of developing certain types of cancers. A minimum 5 g soy protein per serving is recommended along with necessary FDA criteria for the use of a health claim. The FDA will rule on this petition shortly.

Categories of Ready-To-Drink Soy Protein Nutritional Beverages

Beverages containing soy protein can be classified according to their position in the marketplace:

- *Milk-plus* — These beverages contain cow's milk plus soy protein. Historically, they were milk-based beverages to which soy protein was added as an economical protein source; in many instances, soy protein received little recognition on the labels. Today, this practice is changing as dairy companies look to gain an advantage from the health benefits of soy protein. These products are marketed to the mainstream consumer.

- *Milk alternatives* — These beverages contain either no dairy ingredients or at least no lactose, yet they are formulated to approximate the composition of cow's milk. The nutrient profile will include vitamin and mineral fortification. These are marketed to lactose-intolerant consumers who want a nutritious beverage to replace common uses of cow's milk.

- *Soymilks* — These beverages are based on all-vegetable ingredients, with soy as the sole protein source. The nutrient profile generally does not match that of cow's milk. These beverages range quite widely in flavor, texture, and color due to varying processes and ingredients. They have a growing market that includes specific ethnic and religious groups, vegetarians, and health-conscious consumers and have become a widening part of the mainstream consumer market. Some countries do not allow use of the term "milk" to describe these products. In such cases, the products are marketed as soy beverages or under branded names.

- *Pharmaceutical/nutritional* — These beverages are scientifically formulated and clinically tested for specific human diseases or developmental states. Some examples would be infant formula, adult nutritional supplements, and enteral feeding formula. Depending on the country, these are frequently sold via medical referral within pharmacies.

- *Meal replacers/weight loss* — These are beverages formulated to provide balanced nutrition as part of an overall diet plan. Typically, each serving of this beverage will deliver one quarter to one third of a person's requirements for essential vitamins and minerals. Nutrients delivered per calorie of beverage are another important characteristic. The market for these beverages is a person struggling with obesity or attempting to remove last winter's fat layer for cosmetic reasons.

- *Cream alternatives* — Coffee whiteners and whipped toppings are examples of cream alternatives. Food manufacturers originally developed formulated dairy-like cream alternatives with casein as the protein source and vegetable fat as economic and functional replacements. Further economic restrictions and the growth of niche markets with religious, ethnic, and dietary needs demanded replacement of the casein. These products are occasionally consumed alone but are more likely to be consumed as part of another product, such as within or on top of a coffee beverage.

- *Fortified juice* — This is a refreshing fruit-flavored drink with soy protein included (<1.5%) for nutrient fortification. The product viscosity and appearance should be similar to juice that is not fortified. The product may contain real juice or not. Frequently, other nutrients may be fortified within the same product, such as vitamin C and calcium. The market for this product is children, women, and health-conscious adults.

- *Fruit smoothies* — Fruit smoothies are another version of a fruit-type drink but with a higher protein addition (1.5 to 3%). These beverages are thicker in texture than juice and opaque in appearance. They may or may not have any actual fruit. This category has two subsets: One group resembles drinkable yogurts and the other recreates the blended fruit shakes that have been popularized at smoothie bars. These beverages are targeted at health-conscious and active adults seeking added energy and more nutrition in tasty, fruity flavors.

For the remainder of this chapter, beverages will be considered from a physical, functional, and chemical position. Although we may provide a formulation for a specific market, sample formulas have been selected to illustrate differing physical, functional, and chemical situations. Generally speaking beverages are produced at a pH near neutral (6.5 to 7.5) or at a low pH (3.5 to 4.5). These differing pH conditions have a significant impact on protein functionality in beverages. In beverage shorthand, these are considered as ready-to-drink acid (RTD-A) and ready-to-drink neutral (RTD-N) products. Fruit smoothies and protein-fortified fruit juices are examples of RTD-A beverages; soymilk, infant formula, and cream alternatives are examples of RTD-N products.

Soy Proteins

Soy Protein Classification

Several ingredients can potentially be utilized as the raw material source for soy protein in nutritional beverages. It is important to recognize that, although all of these raw materials may be accurately ascribed as soy protein in an ingredient label, significant differences exist in the composition, beverage functionality, and nutritional attributes of each raw material. These differences are derived from agronomic factors, bean varieties, degree and type of bean fractionation processing, and post-isolation processing.

The native soybean contains approximately 40% protein with the balance being composed of oil, fiber, soluble sugars, and hulls. This protein in the bean is a heterogeneous mixture of storage proteins and metabolic proteins (enzymes and structural proteins). One approach researchers have used to fraction and classify these proteins is based on their solubility characteristics. One class of proteins, albumins, is soluble in water and another class, globulins, is soluble in a salt solution. Soybean proteins are predominantly globulins. The globulins can be further fractionated, again by solubility, into glycinins and conglycinins (also known as vicilins and legumins). The glycinins have less solubility in a salt solution than conglycinins.[24] Another classification system for soybean proteins is based on their centrifugal sedimentation characteristics in sucrose gradients. According to this methodology, the bean has four major protein components, described as 15S, 11S, 7S, and 2S fractions. Glycinins are predominantly in the 11S fraction and conglycinins are predominantly in the 7S fraction.[38]

Under normal conditions, glycinin and conglycinin exist as heterogeneous protein polymer complexes formed from multiple subunits with a molecular mass of 360 kDa for glycinin and 180 kDa for conglycinin.[17,45] Soy proteins may further form larger aggregates as a result of different types of commercial processing, or the aggregates may be broken down by additive introduction. Detailed discussions on this topic are available elsewhere.[17,21,42] The ratio of glycinin to conglycinin in a protein product varies depending on the soybean type and processing conditions. Furthermore, the ratio of glycinin (11S) and conglycinin (7S) can affect the functional and nutritional properties of the protein product.

Of the four major categories of protein sources for beverages, three are protein products, where protein has been isolated and fractionated from other parts of the seed. These categories are soy flour, soy protein concentrate, and isolated soy protein. Another potential source is the straightforward aqueous extract of soybeans. For this text, the latter source will be called whole-bean extract (WBE). In most cases, WBE is utilized as prepared in a liquid, diluted form. No fractionation of the component proteins is performed, although physical and thermal processing conditions may modify their native characteristics. The protein content of the dry matter will be 42% (see Table 11.2), although as used the protein content is less than 4%. WBE is the source for protein in traditional soymilk. The technology for traditional soymilk is reviewed in a variety of sources.[4,22,35,44]

Soy Protein Product Manufacture and Chemical Composition

Soy protein products fall into three groups based on protein content. All three basic soy protein product groups are derived from defatted flakes (see Figure 11.1). Soybean (defatted) flours are the soybeans from which hulls

TABLE 11.2

Composition of Some Commercial Products Derived from Soybeans

Component	WBE (%)[a]	Flour (%)[b]	Concentrate (%)[b]	Isolate (%)[b]
Protein ($N \times 6.25$)	42	56–59	65–72	90–92
Fat	23	0.5–1.1	0.5–1.0	0.5–1.0
Carbohydrates	22	32–34	20–22	3–4
Ash	5.2	5.4–6.5	4.0–6.5	4.0–5.0
Fiber	15	19–20	16–26	<0.2

[a] Whole bean extract; a typical liquid form of WBE contains 91% water (Solae LLC data).
[b] Data from Endres.[7]

and oils have been removed; soy protein concentrates are defatted flour from which sugars and oligosaccharides have been removed; and soy protein isolates are defatted flour from which fiber and acid-soluble proteins, sugars, and oligosaccharides have been removed.[7] The products have a minimum of 50, 70, and 90% protein on a dry basis, respectively, as a result of these fractionations (see Table 11.2).

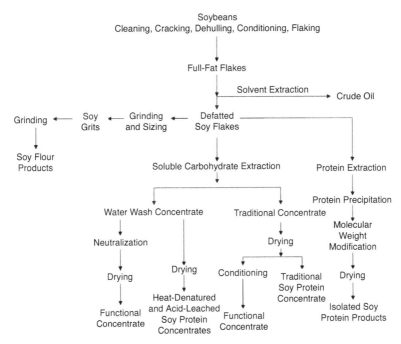

FIGURE 11.1
Soybean processing pathways.

Soy Flour

Soy flour (SF) is made by grinding and screening soybean flakes either before or after removal of oil. They contain about 40 to 54% protein on a dry basis, depending on the oil removal. SF can be purchased with varying fat content, particle size, and degree of heat treatment. For most beverages, a low fat content (<1%), small particle size (<10 μm), and low heat treatment (>85 nitrogen solubility index [NSI]) are preferred. One advantage of full-fat SF is the potential to deliver organic certification status for a beverage. Several finely ground organic SF products and whole-bean powders have been introduced in recent years. Still, SF is rarely used in formulated RTD beverages due to its limited protein content, poor flavor, low solubility, and rough mouthfeel; however, it is used by some soymilk manufacturers and in low-cost beverages.

Soy Protein Concentrates

Soy protein concentrates (SPCs) are prepared by four different processes: acid leaching (at pH 4.5), extracting with aqueous alcohol (60 to 90%), denaturing the protein with moist heat before extraction with water, and size exclusion separation by membranes. This processing removes soluble sugars, certain bean flavors, antinutritional factors, and enzymes that can cause off-flavors; however, except for the membrane-processed protein, the resultant proteins do not have good solubility properties due to protein aggregation during isolation. The acid-leached protein can be neutralized to improve solubility above 70 NSI, but the flavor of this product has not been widely accepted. The low-water-soluble (aqueous alcohol extraction) SPCs can be subjected to heat (steam injection or jet cooking) and physical treatment (homogenization) to increase solubility. These latter two forms of SPCs are known as functional concentrates.[7] Recently, a new patented SPC that uses membrane separation technology has become available on the market.[37] This process involves water extraction of defatted soy flour, centrifugal separation of fiber, membrane separation of larger molecular size proteins from assorted small-molecular-weight organic materials, and drying. Compared to the SPCs produced by traditional processing methods, membrane-processed soy protein concentrates offer improved solubility, emulsification, flavor, and naturally occurring isoflavones.

Soy Protein Isolates

Soy protein isolates (SPIs) are the most highly refined soy protein products commercially available. The primary process for SPIs involves alkaline, aqueous extraction of defatted soy flakes; centrifugal separation of fiber; acidification to the protein's isoelectric point (pH 4.5); separation of insoluble globulins from acid-soluble components; pH neutralization; and drying.[7,43] A variety of isolated soy protein products based on this process that offer different solubility, emulsification, and viscosity properties for beverages is commercially available. SPIs offer more versatility as a protein source than

TABLE 11.3

Relationships between Physical Properties of Proteins and Beverage Attributes

Physical Property	Beverage Functional Attribute
Solubility	Appearance, mouthfeel, sediment, suspension stability
Emulsification	Suspension stability, mouthfeel, appearance, color
Viscosity	Mouthfeel, stability, flavor
Flavor binding	Flavor
Particle size	Mouthfeel, color, and appearance
Heat stability	Color, suspension

other soy protein products for beverages. From a beverage manufacturer's standpoint, SPIs are easy to use and require less processing input to obtain stability and safety. The high protein content of SPIs allows for formulation flexibility, especially to create unique formulated beverages such as low-carbohydrate, high-protein, and low-fat products. SPIs provide better taste and mouthfeel in beverages and reduce the need for masking agents. Specialty SPIs have been developed that can be used in the production of emulsified products such as nondairy coffee whiteners and whipped toppings.

Physical Properties and Functionality

Whenever proteins are a major component of RTD beverages, they will affect the sensory attributes. Although the relationship is not exact, it is safe to say that the physical properties of proteins are related to certain functionality characteristics in beverages. For example, proteins that exhibit good solubility will produce better beverage mouthfeel and suspension stability. The relationships between protein physical properties and beverage functional attributes are summarized in Table 11.3. More detailed discussion regarding protein functionality can be found in other publications.[17,42]

The physical properties across the range of soy protein products are quite different, due to the differences in component composition (protein, fat, and fiber content), preparation, and processing. Also, the physical properties of proteins within the same group (for example among SPIs) can differ from each other to a great extent (see Figure 11.2). Beverage developers should consult their protein suppliers to determine the appropriate soy protein product. In general, good functional proteins in beverage system are those with high solubility, high emulsification, and proper viscosity for the targeted market. Proteins with high foaming and high gelling properties would have a negative impact on beverage quality. Most soy protein products are delivered in dry powder form and must be rehydrated when used; even the ideal protein from a physical property standpoint can produce poor beverage results if not handled correctly.

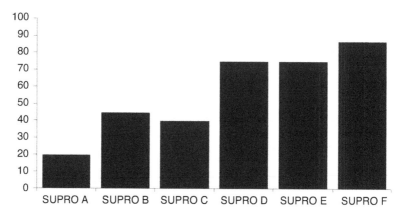

Nitrogen solubility index (%) of various soy protein isolates. (Data courtesy of Solae LLC, St. Louis, MO.)

Key Factors for Beverage Quality

Dispersion and Hydration of Dry Protein Products

Two very important processes must be understood to prepare good-quality beverages containing soy protein products: protein dispersion and hydration. Protein dispersion is the mechanism of introducing powder into a liquid that results in homogeneous slurry without lumps. The process must break up the protein into individual spray-dried particles and induce water uptake by the protein to gain optimal functionality. Hydration refers to the state of water association with the hydrophilic groups of hydrocolloids such as proteins, gums, and starch. The beverage manufacturer should strive to achieve maximum incorporation of water into all areas of the hydrocolloid polymer, as this is important to achieve maximum functionality in the liquid system. Figure 11.3. shows how a well-dispersed and hydrated protein solution results from correct processing. The three stages of dispersion and hydration are identified (i.e., dispersing and wetting, swelling, and dissolution) to illustrate the movement from dry powder to a homogeneous solution. During processing, the beverage processor must set the appropriate conditions to achieve this transition. Lumpy product, excessive sediment, or poor emulsions can be caused by improper dispersion and hydration.

Factors Affecting Hydration

Time, temperature, shear energy, pH, and ionic environment are five factors affecting the hydration of soy protein products. At a constant water temperature and low shear environment, prolonged hydration time (beyond 2

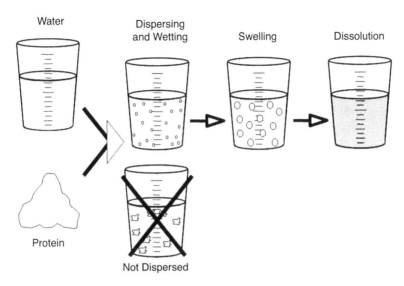

FIGURE 11.3
Physiochemical properties of protein hydration.

hours) results in good protein hydration. These conditions are generally not acceptable to modern processors. When the temperature is increased to 82°C (180°F), proper hydration is reached within a relatively shorter mixing time, 20 to 30 minutes. Higher shear energy (say, in the form of a shear pump or colloid mill) also facilitates hydration and can shorten the hydration time.

Soy protein products are amphoteric substances, which means that they ionize as both acids and bases in aqueous solution. This property is due to the fact their constituent amino acids contain at least one carboxyl and amino group. Protein functionality is affected in total by the sum of the charges across the peptide polymer. The net charge for the proteins in soy protein products such as isolated soy protein is zero around pH 4.5 (see Figure 11.6). This is also the point of minimum water solubility. As the pH decreases below 4.5 or increases above 4.5, the protein solubility increases to a maximum around 2.8 and 7.2, respectively. It is virtually impossible to fully hydrate soy protein products if the attempt is made in a liquid with suboptimal pH conditions. For most beverages, hydration at a neutral pH around 7.0 is recommended.

High ionic strength in the water can also reduce the solubility of soy proteins. Tests have shown that a sodium chloride concentration around 0.1 M significantly decreases the nitrogen solubility index (a measure of protein solubility) of soy protein isolates. It is preferable to hydrate soy protein products in a low ionic environment (<0.05 M as sodium chloride). When the soy protein products have been hydrated, additional ionic materials (monovalent types) can be added without altering the functionality of the protein.

Due to the same chemistry that creates amphoteric properties, soy proteins can also react with certain minerals in ways that prevent hydration. Divalent cations such as calcium and magnesium link to two negatively charged carboxyl residues of the protein. These links occur inter- or intramolecularly; the result is protein aggregation. For this reason, water hardness must be understood and properly managed for protein hydration. The use of water-softening and chelating agents such as citrates and phosphates will make these ions unavailable to interact with the protein products.[26]

Recommended Temperatures

Several systems can properly disperse soy protein products. Powder funnels with recycle systems or industrial blenders with recycle have been most successfully utilized. The protein should be dispersed into water at 20 to 60°C (70 to 140°F), preferably 40 to 55°C (110 to 130°F). The minimal amount of shear effect to obtain dispersion should be used, and shearing should not be continued after the powder has dispersed, as this may result in air incorporation. After increasing the water temperature to 72 to 82°C (170 to 180°F), mixing continues for 15 minutes.[26] At this point, the protein should be hydrated. Some processors prefer to homogenize at about 17.0 MPa (170 bar; 2500 lb/in.^{2}) also, to ensure full hydration.

Formulation and Selection of Ingredients

Protein Attributes

A typical RTD beverage contains about 90% water; thus, the ability of the protein to interact with water and other ingredients such as oil, minerals, and carbohydrate determines the beverage properties and its stability. Depending on its type and concentration, protein can greatly affect the overall properties of the beverage, including viscosity, mouthfeel, stability, flavor, and appearance.

Effect of Sugars on Browning

When carbohydrates are used together with proteins, the choices of proteins and carbohydrates can significantly determine the final color of the finished beverages. One should select carbohydrates that will deliver the correct sweetness for consumer hedonics while not causing too much browning during heat processing. Generally, the rule is to avoid reducing sugars, unless the browning is a required flavor characteristic (e.g., chocolate or caramel). Fructose is a good example of a reducing sugar. In Figure 11.4, several different protein sources have been mixed in a beverage model that contains

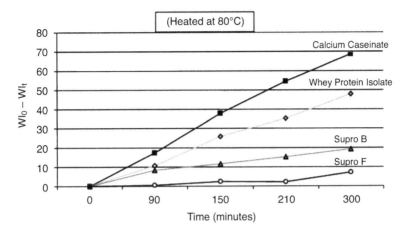

FIGURE 11.4
Maillard reaction of milk proteins or isolated soy proteins with fructose. WI is the whiteness index $(L - 3b)$.

fructose as the sole carbohydrate. Each beverage was heated at 80°C for different times to mimic color generation during processing. In the figure, the color change (as measured by the difference between a starting sample and samples heated for different times) is provided in Hunter Laboratory whiteness index (WI) units. The figure demonstrates that the SPI products had fewer chemical interactions with the fructose, resulting in whiter finished beverages.[16]

Hydrocolloid Stabilizers

Gum stabilizers are commonly used to help suspend the protein, minerals, and fat emulsion in beverages. They also build body and mouthfeel. The primary gums are carrageenans, colloidal microcrystalline cellulose gel, cellulose gum, pectins, and sodium alginates. Gums build intermolecular bonds through electrostatic linkage, creating interstitial matrices. This prevents protein movement in the fluid and increases the viscosity of the beverage to slow down gravitational effects.[26]

Emulsifiers

Soy proteins have good emulsifying properties; however, most beverage processors will add additional emulsifiers to increase the rate of emulsion formation and improve the fat or oil suspension through smaller fat globule sizes. Emulsifiers reduce the interfacial tension between fat or oil with water. Mono- and diglycerides and lecithin are emulsifiers commonly used in RTD soy beverages. Emulsifiers should be premelted and distributed in the fat or oil prior to addition to the aqueous beverage.

Selection of a Buffering System

As indicated earlier in the section on protein hydration, the presence of divalent cations, such as calcium and magnesium, in the process water and other ingredients can have negative impacts on beverage quality. One way to deal with this issue is to determine experimentally the quantities of chelating (or sequestering) compounds necessary for the beverage to be stable. A calcium ion electrode can be used to monitor the available ions in a beverage during makeup, then the amount of citrates or phosphates required for subsequent beverages can be predicted. The sources of sequestering agents can be selected from sodium citrate, potassium citrate, sodium hexametaphosphate, or potassium phosphate (dibasic).

Flavoring

Flavorings affect the overall flavor profile of liquid beverages more than any other factor. Artificial flavors work best in RTD beverages because they withstand heat processes better than natural flavors. For labeling purposes, natural and artificial flavors may often be the best alternative. Certain flavors work better with soy than others. For "brown" flavors, hazelnut, chocolate, English toffee, coffee latte, caramel, rum, raisin, and malt work well. Other flavors that work with soy proteins are cream, almond, passion fruit, peach, apple, cereal, coconut, banana, and honeydew melon.[26]

Colorants

The overall appeal of a beverage can be improved or destroyed by the color selection. As with flavors, artificial colors will usually provide better heat processing stability than natural colors. Some important colors used in soy beverages are titanium dioxide (white), Red No. 3 (watermelon red), Red No. 40 (orange red), Blue No. 1 (turquoise blue), and Yellow No. 5 (lemon yellow). Caramel colors (tans and brown) are one exception to the natural rule, although some do not consider them to be natural. In some cases, color fading will occur in beverages containing ascorbic acid as a vitamin or oxygen scavenger; oxygen and light exposure should be prevented during processing or after packaging to minimize color fading.

Fat Sources

In general, fat or oil sources should be selected that provide good flavor stability. Such key parameters as peroxide values that are <1 mEq/kg fat, acid values that are <0.1 mg KOH per g fat, and a specification for oxidative stability are required. Increasingly, however, certain market segments are requesting oils with special nutraceutical benefits, such as eicospentanenoic acid (EPA) and docosahexaenoic acid (DHA). These fatty acids from fish oil are very prone to oxidative deterioration, so special steps must be taken to stabilize them with antioxidants, metal chelators, and emulsifying agents.

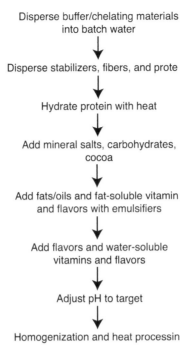

FIGURE 11.5
Process flow for typical RTD beverage.

Processing

Beverage Makeup

Usually, six to seven steps are required for the makeup of beverages. Makeup is the preparation of the liquid mix prior to thermal treatments. These steps are outlined in Figure 11.5.

Thermal Treatments

The heat processing steps in beverage manufacturing are critical control points for preventing pathogens and food spoilage organisms. The selection of appropriate process controls for heat processing must take into account the beverage type, microbial loading of raw materials, finished product packaging, and intended distribution. It is recommended that product developers work with their process authorities to determine these conditions. Table 11.4. presents some conditions used for RTD-N and RTD-A products. Because RTD beverages must be given sufficient heat treatment

TABLE 11.4

Examples of Heat Process Conditions for Beverages

Heat Processing	Batch Type	High-Temperature, Short-Time (HTST)	Flash
Pasteurization	63°C for 20–30 minutes	72°C for 15 seconds	85–95°C for 5 seconds to 2 minutes
Ultra-high temperature	Not applicable	135–150°C for 2–12 seconds	140–150°C for 2–5 seconds
Retort sterilization	115–120°C for 20–30 minutes	Not applicable	Not applicable

to achieve a designed shelf-life, the potential formulations may be limited by this process step. Higher and longer thermal treatments accelerate the interactions between soy protein molecules, water, and other ingredients (minerals, for example) in the beverages. Also, the maintenance of good flavor becomes more problematic in excessively heat processed beverages. Because many soy beverages are sold in shelf-stable packaging, the heat processing creates a product development challenge for optimal taste and stability.

Handling (Foaming)

Foaming is a defect in processing that must be avoided, as it can cause product losses and prevent homogenization and heat processors from providing adequate treatments. It can prevent fillers from running at normal rates or, worse, result in underfilled product. Some key steps can be taken to avoid foam. The process system should be checked for any transfer spots that cause fluid to drop from heights into tanks unguided. Tanks should be mounted with baffles to prevent vortexes from forming during mixing. When dry products are dispersed, high-speed mixers should be decreased to speeds that enable fluid movement but not air incorporation. Adding oils or fats to the protein hydration water will reduce surface activity and foam stability. Approved antifoams should be used at recommended rates. Approved defoamers should be available at critical process steps.

Beverage Applications for Soy Protein

Ready-To-Drink Neutral pH

As discussed earlier in this chapter, beverages can be categorized by a number of characteristics. All beverages fall into two main groups — those at low pH (RTD-A) and those near neutral pH (RTD-N). This section deals with three types of RTD-N beverages. These different types are presented to expose the reader to the key attributes and issues for consideration in developing and producing RTD-N beverages with soy protein products. Products containing significant dairy proteins, all soy protein nutritional beverages, and high-fat beverages are considered. The information should allow the reader to adapt solutions for products like those presented or ones that may require combinations of the technological recommendations. For example, the section on high-fat products would be useful to anyone experiencing emulsion problems.

Beverages Containing Dairy Proteins

Beverages containing dairy proteins dominate consumer markets in most Western countries. Given the economics and health benefits of soybean protein, increasing efforts have been made to combine the proteins. One can find milk-plus, recombined sterilized milk, adult nutritional, sports performance, and weight management beverages on the store shelves today. Compositionally, these may range in protein content from 3 to 5%. Dairy drinks must be stable through the selected heat processing. The product should still be of good quality after a 4- to 6-month shelf life at ambient temperatures. No bacterial spores should survive heat processing to cause product instability during storage. Any fat must stay homogeneously suspended. The protein and minerals must not precipitate and form an insoluble layer that cannot be resuspended during product use by the customer. The color and flavor of the product must be characteristic of the flavor variety intended, although typically the product is slightly darker and stronger in flavor due to the heat processing. The product viscosity must be low for easy drinking.

Optimal product quality will require good solubilization of protein sources. It is recommended that the soy protein be hydrated into water rather than milk or liquid dairy whey. If dry whey is to be added, it should not be added until after protein hydration has occurred. After protein hydration, fat should be homogenized into the hydrated proteins at a minimum of 17.5 MPa (175 bar 2500 psi). Water hardness should be <100 mg/kg total hardness as calcium carbonate. The least sedimentation and best color are achieved through the use of maltodextrin-type or corn-syrup-solids carbohydrates; however, sweet dairy whey typically provides the best economics as a carbohydrate source. In order to arrive at the same composition as milk, 0.313 kg of isolated soy proteins plus 0.687 kg of sweet dairy whey solids should be used to replace every kilogram of skim milk solids. If it is desired to obtain the same calcium per kilogram of blend as for skim milk powder, an additional 0.03 kg of tricalcium phosphate would have to be added.[15] The addition of sucrose (1 to 3%) will reduce the drying perception (astringency) sometimes observed in such beverages. Certain stabilizers can improve product suspension stability; their use is particularly important with chocolate beverages. Kappa carrageenans are frequently used with dairy proteins, but iota and lambda

TABLE 11.5

Formulas for Ready-To-Drink Neutral Beverages

Ingredients	Milk-Plus (%)[a]	Nutritional Beverage (%)[b]	Coffee Whitener (%)[c]
Tap water	17.540	89.549	73.288
Buffering blend	—	0.300	0.200
Stabilizer blend	—	0.400	—
Isolated soy protein and stabilized calcium	—	3.810	—
Isolated soy protein	0.560	—	0.800
Whole milk (12.8% solids)	80.000	—	—
Vegetable oil	0.630	1.050	
Vegetable fat	—	—	10.100
Emulsifier blend	0.020	—	0.600
Sucrose	—	4.800	—
Corn syrup solids	—	—	15.000
Dairy whey	1.200	—	—
Sodium chloride	—	0.030	—
Tricalcium phosphate	0.050	—	—
Vitamin/trace mineral premix	—	0.011	—
Vanilla flavor	—	0.050	—
Butter flavor	—	—	0.012
Total	100.000	100.000	100.000

[a] Data from Gottemoller.[15]
[b] Solae LLC, unpublished data.
[c] Data from Kolar et al.[20]

carrageenans have been shown to work better with soy proteins. The addition of cream-type flavors along with other flavors (e.g., chocolate, strawberry) increases the hedonic acceptance. Historically, fat sources have mimicked the melting properties of milk fat. Today, with the number of lower fat products in the market, consumer mouthfeel expectations have changed. It is more likely that highly nutritious oils are used. The melting characteristics have become secondary criteria.

A formula for a milk-plus beverage is provided in Table 11.5. The basic process is to follow the scheme provided in Figure 11.5. This formula will work well for a pasteurized product. If the product is to be retort sterilized or ultra-high-temperature (UHT) treatment and aseptic packaged, additional stabilizers should be added.

All-Soy-Protein Nutritional Beverages

Many beverages today rely on soy protein as the only protein source. Soy-based infant formula, age- and disease-specific adult nutritional beverages, milk alternatives, weight management and meal replacement beverages, and formulated soymilk are examples. The protein content can range from 2 to 5% and may come from any of the soy protein products mentioned, although soy concentrates and soy protein isolates are used most often.[7] Soymilk from whole-bean extract would also fall into this category, although its taste is not appreciated in all Western markets. For example, when Australian commercial samples of formulated soymilk containing soy protein isolates were compared to whole-bean extract soymilk, the formulated milks provided better taste and aroma than WBE soymilk.[12]

The products generally contain vegetable fat or oil (3%), carbohydrates, sequestering or buffering agents, stabilizing hydrocolloids, emulsifiers, salt, and flavors. The selection and importance of these ingredients were discussed earlier in this chapter. The soy protein beverages may contain some amount of fortification which will be determined by the target market. The fortification targets must be evaluated for impact on beverage quality. Color, flavor, and suspension stability are attributes that can be negatively impacted by improper selection of fortification ingredients.

In general, low viscosity is desired for these beverage products. Higher hedonics are achieved for lower viscosity beverages unless they are market positioned as shake- or meal-type products, in which case a thick and full mouthfeel is preferred. Beverage viscosity will be determined by the choice of protein, carbohydrates, and hydrocolloid stabilizer. Higher viscosity will improve suspension stability of the beverages. In many cases, the beverage developer must balance the beverage viscosity (causing a specific hedonic response) against the required storage stability.

Table 11.5 presents a formula for a soy nutritional beverage. The process follows the steps outlined previously in Figure 11.5. The protein in this formula contains a soy protein isolate with stabilized calcium. This patented technology enables the easy incorporation of colloidal calcium for nutritional fortification. The beverage pH should be 7.0 to 7.2 at the end of mix preparation.[23] This formula would be best heat processed under UHT conditions (see Table 11.4).

High-Fat Beverages

Certain beverages require significant amounts of fat or oil to deliver the functional performance and nutritional composition expected by the consumer. Some examples are coffee whiteners (creamers), enteral feeding formulas, and whipped toppings. A primary function of protein is to assist in the formation and stabilization of oil-in-water emulsions. Practical examples of these are nondairy coffee creamers, whipped toppings, and pharmaceutical enteral formulas for various disease-stressed needs. These beverages may range in protein from <0.5 to 4%. The coffee creamers and whipped topping would represent beverages with low protein levels, just enough to stabilize the emulsion. As the ratio of protein to oil increases, emulsion stabilization is generally easier. Soy protein isolates are the primary protein choice for such beverages.[7]

Soy protein isolates are not a "drop-in" replacement for casein or caseinate in coffee creamers; the emulsifier system may require changes. Emulsifiers having a higher hydrophilic-to-lipophilic balance (HLB) have been found to perform better in high-fat emulsions with soy protein products. Typically mono- and diglycerides are used to assist emulsion formation with caseinates; however, these emulsifiers may not always provide optimal performance in high-fat emulsions containing soy protein products. Diacetyl tartaric acid esters of mono- and diglyceride (DATEM), sodium stearoyl-2-lactylate, and polysorbate 60 are examples of emulsifiers having an HLB of 9 or greater and have been shown to perform well in coffee creamers containing soy protein products.[20] Local regulations must be consulted regarding currently acceptable emulsifiers, as some countries have restricted the application of some of them.

The method of processing will have an effect on the performance of coffee creamers. Homogenization is important because it assists in the preparation of an emulsion having fat globules 0.2 to 0.4 μm in diameter. Good homogenization results in uniform fat distribution, improved protein hydration, smoother and creamier body, and whiter color. Rapid cooling of the emulsion immediately after homogenization assists in producing a stable emulsion.[20] The effectiveness of homogenization varies with mechanical conditions of the homogenizer, even though the same pressure may be used. Homogenizers should be frequently inspected and maintained for seal, valve, and piston wear.

Different flavoring systems may be required. Coloring agents such as titanium dioxide may have to be added to achieve whitening power equal to that of caseinate-based formulas. In coffee creamers, the product must also be stable in the acid environment of coffee or tea. In addition, the hardness of water in the coffee should not destroy the whitening power of the coffee creamer nor cause emulsion breakdown. In order to achieve this, the formulation must contain sufficient chelation compounds and buffering agents. The protein product should be selected for good acid solubility and emulsification; it should not be negatively impacted by divalent cations up to 200 mg/kg as calcium carbonate.

Table 11.5 presents a formulation for a liquid coffee whitener. For this formulation, the buffering salts are added to the water and allowed to solubilize as the water is heated to 50°C. The soy protein product is then hydrated as mentioned earlier. Following hydration, hydrophilic emulsifiers are added to the protein. The lipophillic emulsifiers are added to the fat and mixed until fully dispersed (this may require heating the fat up to 60°C). The fat is then added into the protein, then the corn syrup solids are added and mixed until a uniform suspension is obtain. The product is then pumped to heat processing at 139°C for 10 seconds and cooled to 82°C. The product next goes to homogenization in two stages at 2000 and 500 psi (141 and 35 bar) and is then cooled to <27°C. Such a product is then aseptically packaged into sterile containers.

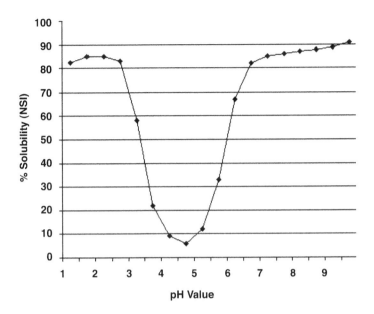

FIGURE 11.6
pH/solubility curve for the main soy proteins. (Data courtesy of Solae LLC, St. Louis, MO.)

Ready-To-Drink Acid

Protein-containing RTD-A beverages can be divided into two types based on viscosity in the finished product. The first type is juice like with very low viscosity, whereas the second can be described as a smoothie. Usually this type of product has a higher viscosity and a creamier mouthfeel. By definition these products contain a high acid content (i.e., have a low pH), usually with a pH in the range of 3.6 to 4.2. In this pH range, soy protein possesses a low solubility (Figure 11.6); therefore, it is a challenge to make stable acidic RTD beverages. The beverage acidity may come from natural juice content or may be the result of direct acidification. Any food-grade acid could be used for this latter purpose; however, the most commonly used types are phosphoric, citric, and lactic acids. A precise finished beverage pH can be achieved through the selection of an appropriate buffer system. Acidification could also be accomplished slowly using fermentation with suitable lactic cultures, or glucono-delta-lactone (GDL). Common fruit juices used are white grape and apple. Orange juice is rarely used, as it contains pulp that may lead to accelerated sedimentation.

Stabilization

The isolelectric precipitation of soy proteins has been widely studied. These studies date back many years.[18,30] Isoelectric precipitation of soy proteins at pH 4.5 leads to the formation of small primary particles that aggregate to form much larger particles. This aggregation has to be inhibited in order for the protein particles to remain suspended. A similar phenomenon has long been observed in dairy systems.[13] The use of acid polysaccharides prevents the destabilization of acid milks. Glahn[13] reported that pectin, propylene glycol alginate (PGA), and sodium carboxymethylcellulose (CMC) all successfully stabilized acidify milks. He attributed this success to a coulombic repulsion between the particles after the casein in milk has acquired a negative charge due to its acidification. Soybean hemicelluloses have recently been added to the list of successful stabilizers in acid beverages.[25]

Electrostatic and hydrophobic forces are responsible for binding the stabilizers onto the proteins. The same phenomenon was reported for pectin with soy proteins.[14,19,30] Klavons et al.[19] reported that methoxylated pectin prevented the agglomeration of the primary particles produced by isoelectric precipitation and resulted in stable suspensions. The degree of methoxylation was important; fully methoxylated pectin was not as effective in stabilizing the suspension. Although most literature recommends high-methoxy pectins, a recently issued patent claims low-methoxy pectins can also be used.[32]

The molecular weight of high-methoxy pectin is also important, with partially and totally hydrolyzed pectins failing to provide good suspension stability. Dickinson et al.[6] showed that highly methoxylated pectin coaggregates with casein in the interface with fat droplets in casein-stabilized acidic emulsions. Sedlmeyer et al.[34] showed the critical nature of pectin concentration and homogenization conditions in determining the stability of acidified milk drinks.

In acidified milk beverages, the ratio of pectin to protein is critical. As particle sizes are reduced as a consequence of increased homogenization pressure, the requirement for pectin is increased.[34] The same is probably true for soy-containing beverages under acid conditions. Data regarding the importance of degree of shear in process and stability at varying pectin-to-protein ratios do not agree,[14,41] but generally with decreasing soy protein particle size, while holding protein content, the pectin concentration necessary to maintain stability must be increased.[5]

Protein Selection Criteria

Even at low protein contents (<0.5%), the choice of soy protein is important. At higher protein levels, it becomes critical. As noted earlier in the section on RTD-N, manufacturers have built a variety of functional properties into soy protein products. Among these properties are better solubility and heat stability in acid conditions.

Flavors and Colors

These acidified soy protein beverages will have an opaque appearance. The colorant must be selected that will display the desired hue and maintain this during the expected shelf-life. As mentioned earlier, oxygen and vitamin C can be detrimental to color stability. Many different fruit flavors work well with acidic soy protein beverages. Frequently, marketers prefer fruit cocktails (with several fruit flavors) for these beverages, because the consumer is less likely to have a preconceived notion of what the mixed flavor should be so flavors contributed by the protein are less recognizable. It is easier to obtain good consumer flavor acceptance of formulas with fruit juice contents above 10% than below.

Thermal Treatments

Due to the low pH of these products, a heat process to achieve sterility is not necessary, as bacterial spores will not germinate in the pH range being discussed; therefore, a thermal treatment that ensures killing of vegetative bacterial cells and inactivation of any enzymes suffices. This is a much milder treatment than UHT or sterilization conditions. Pasteurization temperatures are typically lower than 100°C with hold times of 5 minute or less;[36] however, as for neutral beverages, the specific conditions must be determined for each beverage based on pH, microbial load and type, shelf-life target, and packaging materials.

Formulation

Table 11.6 provides a formulation for a fortified fruit drink with 3% protein. As described in previous beverages, the protein and stabilizer (pectin) must first be dispersed and hydrated, then the fruit juices, flavors colors, buffers, and acidulants are added. The pH must be checked and adjusted to target prior to homogenization and pasteurization.

TABLE 11.6

Formula for Acidic Beverage

Ingredients	Fortified Juice (%)
Tap water	67.780
Genu® pectin	0.500
Isolated soy protein and stabilized calcium	3.570
Juice (12–15 brix°)	20.000
Sucrose	8.000
Sodium citrate	0.150
Citric acid (50% w/v)	To pH 3.8–4.2
Flavor system	+
Color system	+
Total	100.000

Source: Data from CP Kelco.[5]

Parting Comments

Soy protein processors have developed many beverage-targeted products and application briefs to assist the development of new beverages; however, specific questions can be answered only by the beverage developer to arrive at a "best solution." What type of beverage are you attempting to develop? Have you identified a nutrient or health benefit profile? Do you have limitations on ingredients that can used (this could be for marketing or regulatory reasons)? What type of processing is available? What will be the packaging? Do you have any label requirements? A project will move much more quickly if the developer is prepared with answers to these questions. Soy proteins can be useful ingredients for a beverage developer and manufacturer. Consumers looking for a more nutrition punch from their beverages have found "contains soy protein" a signal to buy. The existing health benefits for soy protein are being augmented by new clinical studies that promise to continue this demand. Technological investments by soy processors have improved the quality of soy protein products, particularly with respect to flavor. Soy proteins can provide the functionality required for fat emulsification and suspension stability. On top of these attributes, soy proteins are economical sources of high-quality protein.

References

1. Anderson J.W., Johnstone B.M., and Cook-Newell M.E., Meta-analysis of the effects of soy protein intake on serum lipids, *N. Engl. J. Med.*, 333, 276–282, 1995.
2. Badger, T.M., Ronis, M.J.J., Hakkah, R., Rowlands, J.C., and Korourian, S., The health consequences of early soy consumption, *J. Nutr.*, 132, 559S–565S, 2002.
3. Bressani, R., World conference on soya processing and utilization, *J. AOCS*, March, 392–400, 1981.
4. Chen, S., Preparation of fluid soymilk, in *Proceedings of the World Congress on Vegetable Protein Utilization in Human Foods and Animal Feedstuffs*, Applewhite, T.H., Ed., AOCS Press, Champaign, IL, 1989, pp. 341–351.
5. CP Kelco, *Application Bulletin: A Taste of Soy*, CP Kelco U.S., Inc., San Diego, CA, October, 2001.
6. Dickinson, E., Semenova, M.G., Antipova, A.S., and Pelan, E.G., Effect of high methoxy-pectin on properties of casein-stabilized emulsions, *Food Hydrocolloids*, 12, 425, 1998.
7. Endres, J.G., *Soy Protein Products: Characteristics, Nutritional Aspects, and Utilization*, Soy Protein Council and AOCS Press, Champaign, IL, 2001.
8. Evans, L., Internal memo, Solae LLC, St. Louis, MO, 2004.
9. FAO/WHO, *Protein Quality Evaluation*, Report of a Joint FAO/WHO Expert Consultation, Food and Agriculture Organization and World Health Organization, Rome, 1990.

10. Fass, P. and Jones, M., The trends and developments in functional beverages, *Soft Drinks Int.*, July, 22–23 2003.
11. Food Nutrition Board, *Recommended Dietary Allowances*, 10th ed., National Academy of Sciences, Washington, D.C., 1989.
12. Ginn, P.W., Hosken, R.W., Cole, S.J., and Ashton, J.F., Physiochemical and sensory evaluation of selected Australian UHT processed soy beverages, *Food Australia*, 50, 347–351, 1998.
13. Glahn, P.-E., Hydrocolloid stabilization of protein suspensions at low pH, *Prog. Food Nutr. Sci.*, 6, 171, 1982.
14. Glahn, P.-E., Personal communication, 1984.
15. Gottemoller, T., *Critical Aspects of Production of Dairy Products with Soy Protein*, presentation to Guatamalan Food Technology Association for the American Soybean Association, Archer Daniels Midland, Decatur, IL, May 31, 1995.
16. Huang, X.L., Impact of Types of Protein on Browning Color Changes in Protein Beverages, paper presented at the 2004 IFT Annual Meeting, Las Vegas, NV, July 12–16, 2004.
17. Kinsella, J.E., Functional properties of soy proteins, *J. AOCS*, 56, 242, 1979.
18. Kiosseoglou, V. and Doxastakis, G., The emulsification properties of soybean protein in the presence of polysaccharides, *Lebens.-Wiss. U.-Technol.*, 21, 33, 1988.
19. Klavons, J.A., Bennett, R.D., and Vannier, S.H., Stable clouding agent from isolated soy protein, *J. Food Sci.*, 57, 945, 1992.
20. Kolar, C.W., Cho, I.C., and Watrous, W.L., Vegetable protein application in yogurt, coffee whiteners, and whip toppings, *J. AOCS*, 56(3), 389–391, 1979.
21. Kolar, C.W., Richert, S.H., Decker, C.D., Steinke, F.H., and Vander Zanden, R.J., Isolated soy protein, in *New Protein Foods*, Vol. 5, Altschul, A.A. and Wilcke, H.L., Eds., Academic Press, New York, 1985, pp. 259–299.
22. Kwok, K.C. and Niranjan, K., Effect of thermal processing on soymilk, *Int. J. Food Sci. Technol.*, 30(3), 263–265, 1995.
23. Lin, S.H.C. and Cho, M.J., Process for the Production of a Mineral Fortified Protein Composition, U.S. Patent No. 4642238, 1986.
24. Liu K.S., *Soybeans Chemistry, Technology, and Utilization*, Aspen, Gaithersburg, MD, 1999, pp. 25–113.
25. Maeda, H., Furuta, H., Yoshida, R., Takahashi, T., and Sato, Y., Water-Soluble Polysaccharide and a Process for Producing the Same, U.S. Patent No. 5710270, 1996.
26. Mai, J. and Lo, G.S., Soy-based nutritional beverages, in *Neutraceutical Beverages Chemistry, Nutrition, and Health Effects*, Shahidi, F. and Weerasinghe, D.K., Eds., American Chemical Society, Washington, D.C., 2004, pp. 266–278.
27. Miller, H.W. and Wen, C.J., Experimental nutrition studies of soymilk in human nutrition, *Chinese Med. J.*, 50, 450–459, 1936.
28. Moore, R.S., An ounce of prevention, in *Chinese Doctor: The Life Story of Henry Willis Miller*, Pacific Press, Mountain View, CA, 1969, pp. 119–128.
29. Mott, S., Soya protein in sports nutrition, *Int. Food Mark. Technol.*, December, 10–13, 1997.
30. Nelson, C.D. and Glatz, C.E., Primary particle formation in protein precipitation, *Biotechnol. Bioeng.*, 27, 1434, 1985.
31. Nestel, P.J., Yamashita, T., Sasahara, T., Pomeroy, S., Dart, A., Komesaroff, P., Owen, A., and Abbey, M., Soy isoflavones improve systemic arterial compliance but not plasma lipids in menopausal and perimenopausal women, *Arterioscler. Thromb. Vasc. Biol.*, 17, 3392–3398, 1997.

32. Patel, G.C., Cipollo, K.L., and Strozier, D.C., Juice and Soy Protein Beverage and Use Thereof, U.S. Patent No. 0104108, 2003.
33. Hughes, K., Ingredient fit for functional foods, *Prepared Foods*, July, 2000 (http://www.preparedfoods.com/CDA/ArticleInformation/features/BNP_Features_ Item/0,1231,114282,00.html).
34. Sedlmeyer, F., Brack, M., Rademacher, B., and Kulozic, U., Effect of protein composition and homogenisation on the stability of acidified milk drinks, *Int. Dairy J.*, 14, 331, 2003.
35. Shurtleff, W. and Aoyagi, A., *Tofu and Soymilk Production*, The Soyfoods Center, Lafayette, CA, 1984.
36. Silva, F.V.M. and Gibbs, P., Target selection in designing pasteurization processes for shelf-stable high-acid fruit products, *CRC Crit. Rev. Food Sci. Nutr.*, 44, 353–360, 2004.
37. Singh, N., Soy Protein Concentrate Having High Isoflavone Content and Process for Its Manufacture, U.S. Patent No. 06818246, 2004.
38. Smith, A.K. and Circle, S.J., *Soybeans: Chemistry and Technology.* Vol. 1. *Proteins*, AVI Publishing, Westport, CT, 1978.
39. *Soyfoods: The U.S. Market 2004*, SPINS/Soyatech, San Francisco, CA, 2004.
40. Steinberg, F.M., Guthrie, N.L., Villablanca, A.C. , Kumar, K., and Murray, M.J., Soy protein with isoflavones has favorable effects on endothelial function that are independent of lipid and antioxidant effects in healthy postmenopausal women, *Am. J. Clin. Nutr.*, 78(1), 123–130, 2003.
41. Tromp, R.H., de Kruif, C.G., van Eijk, M., and Rolin, C., On the mechanism of stabilization of acidified milk drinks by pectin, *Food Hydrocolloids*, 18, 565–572, 2004.
42. Utsumi, S., Matsumura, Y., and Mori, T., Structure–function relationships of soy proteins, in *Food Proteins and Their Applications*, Domodaran, S. and Paraf, A., Eds., Marcel Dekker, New York, 1997, pp. 257–291.
43. Waggle, D.H. and Kolar, C.W., Types of soy protein products, in *Soy Protein and Human Nutrition*, Wilcke, H. L., Hopkins, D. T., and Waggle, D.H., Eds., Academic Press, New York, 1979. pp. 19–51.
44. Wilson, J.C., The commercial utilization of soybeans, soymilk and soymilk derivatives, in *Proceedings of the World Soybean Research Conference*, Buenos Aires, Argentina, March 5–9, 1989, pp. 1750–1766.
45. Yamauchi, F., Yamagishi, T., and Iwabuchi, S., Molecular understanding of heat induced phenomena of soy bean proteins, *Food Rev. Int.*, 7, 283–322, 1991.

12

Soy Product Off-Flavor Generating, Masking, and Flavor Creating

Rongrong Li

CONTENTS

Introduction

Soybean, the hairy-stemmed plant, is one of the stars of the legume family. In Chinese, it is called *da dou*, meaning "big bean." This precious bean was discovered in China more than 3000 years ago. The Chinese, Japanese, and Koreans all claim to have been the first to uncover the rich nutritional value of soy. Soybean is traditionally used in a wide range of East Asian food. From entrée to dessert, from snack to beverage, a variety of foods is made from soybeans. Chinese vegetarian cuisine has incorporated various soybean products to imitate meat texture and appearance for hundreds of years. Soybean curd, also called *tofu* ("bean muscle" in Chinese), plays a major role in these vegetarian foods. Tofu is found in a wide range of products with different textures and flavors. Besides soft-textured tofu, the family also includes partially dehydrated firm and fermented tofu products. The many varieties of tofu can be likened to the cheese family in terms of processing. Tofu processing involves wet milling to produce soymilk, then protein coagulation, dehydration (by compression), fungi fermentation, aging, and so on, depending on the specific requirements of the product. Because traditional fresh tofu is perishable, it normally must be consumed within a day or two. Fungi fermentation and high salt curing are necessary for preservation as well as for flavor development. Aseptic packaging and new thermal processing methods brought shelf-life stable fresh tofu to the market two decades ago. Tofu manufacturers in the United States have made big strides in terms of processing tofu designed for the American palate. The consumer preference for flavored fresh or soft tofu is higher than for plain varieties in the United States.[1]

Soy sauce, a traditional Chinese condiment, is bacterially fermented and has a balanced salty, meaty, savory flavor without much original soybean taste. Today, it has become one of the most popular and well-accepted soy products used in various cuisines around the world. Soymilk used to be an ethnic beverage, as important to some Asian populations as milk is to Western cultures, but the health benefits of soy and innovative new soymilk products have made soymilk more popular throughout the world. Many fruity flavored soy beverages, such as smoothies, and coffee drinks can now be found on the market.

Although the vegetarian market is one of the fastest growing segments in the marketplace today, soymilk and vegetarian meat-like products similar to tofu did not gain popularity in the Western market until about two decades ago, when the health benefits of soy as a regular part of the diet became known to the public. Today, vegetarian imitation meat products, such as vegetarian burgers, meatballs, or chili, are considered mainstream healthy alternatives. These products satisfy consumer craving for meat but eliminate the consumption of saturated animal fat and cholesterols. Soy, wheat, and other textured functional ingredient companies have introduced unique textured proteins and binder systems to create meat-like bases, while the flavor and seasoning industries have created unique flavor systems using only plant-source ingredients to make the vegetarian "meat" products taste as good as real meat products. Soybeans were brought to the United States in the early 19th century. Today, the United States is the world's largest soybean supplier, producing half of the world's total, followed by China and Brazil. Properly processed and refined soybean oil has established itself in the vegetable oil market. Soybean oil contains primarily unsaturated fatty acids (linoleic and linolenic), which are essential to human health; soybean oil, usually sold as vegetable oil, contributes 12% of the energy to the U.S. diet.

Soymilk, tofu, and soy burgers are three major growth areas; according to the United Soybean Board's 2003/2004 annual study, 28% of Americans consumed soy foods or soy beverage once a week or more in 2002 vs. 24% in 1999. Consumer awareness of soymilk increased to 94% in 2004.

The growing awareness of their health benefits and unique functional properties has increased manufacturer demand for high-solubility and better flavored soy products, including soy flour, soy protein concentrates, soy protein isolates, textured soy protein, soy lecithin, and soy isoflavones; however, the off-flavors associated with soy products, generated primarily during processing, have had an effect on the market growth of soy products. Food companies are constantly seeking ways to improve the quality of soy products as well as to create new soy products.

Organoleptic properties encompassing variables such as texture, fibrousness, juiciness, and taste are factors that affect food acceptance. Even a small percentage of flavor ingredients can carry through to the finished products either positively or negatively.[2] The low-cost, high-ingredient functionality and health benefits of soy products have motivated the food industry and researchers to research the causes of off-flavors and to find ways to eliminate them. A deeper understanding of the nature of soy compounds, processing behavior, interactions with flavor compounds and other food ingredients, and stability under processing and storage conditions may help the food industry to identify ways to manipulate off-flavor formation or remove off-flavor compounds. This will lead to an improved overall flavor profile and increased acceptance of soy products as food ingredients.

Identifying the formation of off-flavor compounds during the early stages of soybean processing is essential. Investigating soy product flavors, besides those of the oil, and their impact on overall food flavor can be difficult because off-flavor formation is complicated and food itself is a complex system. Researchers have been investigating the causes of unpleasant off-flavors and seeking ways to improve the flavors for over 40 years. This chapter reviews findings on off-flavor formation in soybean products, soy protein flavor adsorption and binding capacity, possible ways to remove off-flavors, methods of masking off-flavors, and potential ways to make flavored soy proteins.

Soybean Chemical Composition and Soy Products

Soybeans contain about 40% protein, 35% carbohydrates, 20% fat, and 5% ash. Soybean has been an important protein resource for East Asian diets due to its well-balanced amino acid composition, with the exception of sulfur-containing amino acids. The combination of soybean and wheat has been suggested to balance essential amino acids and improve the biovalue of the proteins. Soybean protein contains proteinaceous substances known as trypsin inhibitors, which inhibit the digestion of protein and the nutritionally important hemagglutinins (lectins). Heat inactivation is necessary to eliminate their negative effects on the nutritional quality of soybean protein. Soybeans also contain goitrogens as well as antivitamins to vitamins D, E, and B12. A soy-rich diet requires more calcium, magnesium, zinc, copper, and iron intake.[3] The positive effects of soy protein on lowering cholesterol have been well known for years. A recent study demonstrated that soy protein was inversely associated with total and low-density lipoprotein (LDL) cholesterol concentration without changing the high-density lipoprotein (HDL) level.[4] In general, soybean carbohydrates do not have a bad taste but can cause flatus. The lower human intestine lacks enzymes to hydrolyze raffinose and stachyose. Avoiding direct soybean consumption and using soy protein concentrates and soy protein isolates as food ingredients could remedy such problems.[3]

Soybean oil is composed of triglycerides with varying fatty acids as a part of their structure. The fatty acids in soy oil are primarily unsaturated, oleic (23%), linoleic (51%), and linolenic (7%). It has approximately 15% saturated fatty acids, including lauric, myristic, palmitic, and stearic acids. The high content of the essential fatty acids (linoleic and linolenic) offers great health benefits to consumers; however, unsaturated fatty acids are susceptible to oxidation, both enzymatic and nonenzymatic. Refined soybean oil is a pure oil from which all free fatty acids and other non-oil materials have been removed by chemical and physical means to stabilize oil quality. Soybean lecithin is purified from gums that are the byproducts of degumming during crude soybean oil refining. Commercial unmodified soy lecithin contains three major compounds: phosphatidylcholine, phosphatidylethanolamine, and phosphatidylinositol. It contains approximately 30% linoleic acid and 3% linolenic acid, which are susceptible to oxidation under certain conditions. Soy lecithin is used as an emulsifier in food products and dietary supplements.

Soy flour, soy protein concentrates, soy protein isolates, textured soy protein concentrates, and textured soy protein isolates are popular soy protein products widely used in the food industry. Soy flour is often used in baking to improve texture and bleach the color. Because soy protein concentrates and isolates have a higher protein content, more subtle taste, and better processing functionality than soy flour, they are often used in foods such as beverages, sauces, meats, and soups, as well as baked goods. Textured soy protein products are commonly used as meat replacements in extended meat products and vegetarian meats.

Sources of Unpleasant Flavors Generation in Soy Products

Lipid Oxidation: Autoxidation and Enzyme Hydroperoxidation

Unprocessed soybeans have a very mild flavor and an almost plain taste. Off-flavors are often generated through enzymatic lipid oxidation and autoxidation during soybean processing. Such unpleasant off-flavors restrict the use of soybean products as food ingredients. The enzymes involved in lipid oxidation are identified as lipoxygenases (EC 1.13.11.12). Soybean is rich in lipoxygenases and unsaturated fatty acids. When the beans are intact, lipoxygenase and unsaturated fatty acids are separated by the cell membranes. Direct contact between the enzymes and substrates occurs after harvesting and during the seed-crushing process. Mechanical destruction occurs during dry crushing or wet milling, and the presence of water or heat can accelerate the rate at which lipoxygenases catalyze unsaturated lipid oxidation. Unsaturated, especially polyunsaturated, lipids in soybeans undergo molecular oxygen hydroperoxidation catalyzed by lipoxygenases, at the same time autoxidizing when such mechanical destruction occurs.[5] Three lipoxygenases, L-1, L-2, and L-3, have been identified in native soybean as being responsible for lipid oxidation. Both bound and free linoleic and linolenic acids are major targets of these three lipoxygenases.

The hydroperoxidized lipids are unstable and easily cleaved with or without enzymes, resulting in off-flavor compounds. It has been reported that, during soybean processing, 13-L-c, t-hydroperoxy-*cis*-9, *trans*-11-octadecadienoic acid (13-L-c, t-HPO) generated by lipoxygenases from linolenic is further cleaved by hydroperoxylase, producing *n*-hexanal in the soybean. Hydroperoxylase specifically cleaves 13-L-c, t-HPO and generates *n*-hexanal, while hexanol is produced by linoleic acid 13-hydroperoxide.[6] It has also been reported that *n*-hexanal converts to *n*-hexanol under certain conditions.[7] *n*-Hexanol is considered to be the major source of off-flavors.[6,7] The further cleavage of unstable hydroperoxidized lipids also results in off-flavor compounds such as ketones, aldehydes, furans, alcohols, esters, fatty acids, and amines. Alcohol, short-carbon-chain acids, ketones, and aldehydes are responsible for the typical taste and flavors of soy products. The medium-chain aldehydes pentanal, hexanal, and heptanal were found to be the major class of compounds contributing to the beany, grassy, and stale off-flavors.[8]

Lipoxygenases Activities

Identifying the mechanisms of polyunsaturated fatty acid oxidation is essential to improving the flavor profiles of soy products. Understanding lipoxygenase activities under different conditions will help identify possible ways to avoid or eliminate the generation of off-flavors. Research[9,10] has found that each lipoxygenase has different optimum activity conditions. Understanding the optimum conditions of each isozyme may lead to a way to minimize or inactivate lipoxygenase to improve the soy product flavor quality. For example, the optimum pH for L-1 ranges from 9 to 9.5, and it is less active at neutral pH, while the optimum pH levels for L-2 and L-3 are between 6 and 7. Furuta et al.[9] showed that the optimum temperature range for L-1 and L-3 activity was around 50°C, and activity reduced rapidly with further increases in temperature. The optimum temperature range for L-2 is from 20 to 40°C, and its activity reduces rapidly at higher temperature. Because soybeans are traditionally processed at a neutral pH and at room temperature, it can be concluded that L-2 and L-3 are more responsible for generating off-flavors.[5,6]

Matobo et al.[6] studied a native-type soybean, Suzuyutaka, in comparison with deficient mutant seeds, L-1 null, L-2 null, L-3 null, and L-1/L-3 null, to determine which lipoxygenase isozyme was predominately responsible for *n*-hexanal. The results showed that the L-2 null soybean generated the least amount of *n*-hexanal, and the L-1/L-3 null produced the highest. Furuta et al.[9] also observed that L-2 is most responsible for processing *n*-hexanal. The L-1/L-3 null (L-2 present) soybean was also effective to exogenous linoleic acid, yielding almost three times more *n*-hexanal than the native soybean Suzuyutaka. These data demonstrated that L-2 is most responsible for linoleic acid hydroperoxidation and the production of off-flavor compounds. It has also been observed that L-1 is more stable at 70°C than L-2 and L-3.[6,9] Genetically modified soybeans free of L-2 showed the least accumulation of *n*-hexanal compared to soybeans free of other lipoxygenases.

The thiobarbituric acid (TBA) value and the 1,3-diethyl-2-thiobarbituric acid (DETBA) value are often used to detect lipid oxidation products. The DETBA values and *n*-hexanal levels of two native soybeans, Suzuyutaka and Fukuyutaka, were almost twice those of mutants containing only one of the lipoxygenases.[9] A study of pH dependency showed that native soybeans obtain the highest levels of DETBA at pH 6, L-1 at pH 7 to 9, L-2 at pH 6 to 7, and L-3 at pH 7 to 8, findings similar to those of Axelrod et al.[10]

When substrates are available, enzyme activities depend not only on pH but also on the temperature and time allowed for reactions. Discovering the optimum temperature for lipoxygenase activity is another key to controlling off-flavors. A study on temperature dependency study showed that natural soybean DETBA values increase with a rise in temperature, reaching a maximum at 50°C and decreasing sharply with further increases in temperature. Mutants with L-1 and L-3 both exhibited maximum activity at 50°C, while the mutant with L-2 reached a maximum at about 30°C. Soybean lacking all three lipoxygenases had no response for DETBA within pH 4 to 10 after up to 60 minutes of incubation or increased temperature.[9]

Clean flavor characteristics of soy products are fundamental; furthermore, understanding soy products and flavor compound interactions is critical to gaining greater consumer acceptance of products incorporating soy. The interactions of the chemical flavor compounds with food matrices and individual macromolecules, such as proteins, lipids, and carbohydrates, are the result of physical or chemical entrapment. Covalent, hydrogen, electrostatic, and hydrophobic bonds are common binding forces involved in food system flavor trapping. In soy products, off-flavors generated during early processing are often entrapped or bound by soy proteins, thereby affecting food products during further application and storage.

Oil-Body-Associated Protein and Polar Lipids

The off-flavors (e.g., beany, green, grassy, bitter) found in soy protein products are major technical impediments to the increased use of soy protein in foods. Aqueous extracts of hexane-defatted soy flour at neutral pH and soy protein isolates have an undesirable off-flavor that diminishes their usefulness for incorporation in food materials. Research has shown that polar lipids (PLs) and acid-sensitive polar fractions bind off-flavors and non-protein materials.[11] When PLs such as glycolipids and lecithin were studied in acid-sensitive fractions under neutral pH, they were found to bind to soy products and contribute to off-flavors.[12] Phospholipids have been implicated as a causative substance in oxidative deterioration of food products during processing and storage. Autoxidation of unsaturated fatty acids of phospholipids in commercial oil-free lecithin products produces the most volatile off-flavor compounds. Soy proteins that tend to interact with oil-body membranes and bind to PLs and their derivatives are referred to as oil-body-associated proteins (OBAPs). The OBAPs of 34, 24, 18, and 17 kDa have been found to bind off-flavors.[7] Pure soybean proteins have no aroma and have a neutral flavor. The protein–lipid compounds (OBAPs/PL fractions) usually resist defatting treatment with *n*-hexane. Salting out and conventional centrifugation can be used to remove the fractions and thus prepare soy protein with significantly less volatile off-flavors.

Soy lecithin is another widely used ingredient in food and diet. Unmodified commercial soy lecithin contains three major compounds: phosphatidylcholine, phosphatidylethanolamine, and phosphatidylinositol. These polar phospholipids are the cause of soy lecithin off-flavors and behave in a manner similar to proteins. Phosphatides present in crude soybean oil form sludge during transportation and storage. Hydroscopic phosphatides will absorb moisture from the air, become hydrated, and then settle out, forming troublesome precipitates in the oil. For this reason, degumming treatment is essential to stabilizing soybean oil; also, the degumming process produces soy lecithin. Soybean contains large amounts of phosphatides, the major source of commercial lecithin. Lecithin is widely used in food product applications and nutritional supplements, as well as in other industries, such as feed, chemicals, cosmetics, and pharmaceuticals. Soy lecithin is used as an emulsifier, antioxidant, stabilizer, lubricant, wetting agent, and nutritional supplement.

The functionality of lecithin in various applications is based on the particular system. Although lecithin is an antioxidant in chocolate and margarine, it is susceptible to oxidation in other food systems where water is available. Suriyaphan et al.[2] has investigated off-flavor development in reduced-fat cheddar cheese when unmodified soy lecithin is used as an emulsifier for texture improvement. Soy lecithin was found to give rise to volatile off-flavor chemicals (e.g., alkanals, alkenels, dienals) in low-fat cheeses. Aldehydes such as hexanal; (E)-1,5-octadien-3-one; (E,E)-2,4-decadienal; (E,Z)-2,4-decadienal; (E,E)-2,4-nonadienal; and (E)-2-nonenal had high odor intensity. The off-flavors produced by these compounds are similar to extruded wheat flour off-flavors, as well as other food off-flavors, and they contribute fatty, sweet, corn chip, stale, hay, fried, and bitter notes.[2,13] The two decadienals are the major contributors to off-flavors. No such unpleasant flavors are found in regular cheddar cheese. The common odors of these compounds are green, paint, rancid, and metallic. The flavor of (E)-2,4-nonadienal is fatty and sweet; (E,Z)-2,4-decadienal is corn chip, stale, and hay; (E,E)-2,4-decadienal is fatty and fried; and 2-nonenal is stale and bitter. The decadienals are also the decomposition products of autoxidized linoleic acid. Such compounds have not been detected in cheddar cheese to which lecithin has not been added nor in fat-reduced cheddar cheese containing hydrogenated soy lecithin or oat lecithin. Milk, hydrogenated soy lecithin, and oat lecithin contain very few or no linoleic or linolenic acids. The increased moisture content of low-fat cheddar cheese may accelerate the overall oxidation rate of fat.

Bitter Peptides and Bitter Lipid Compounds

While fresh soybeans have a very mild flavor, bitterness is often detected in processed soy products, primarily due to bitter peptides from soy proteins and products of unsaturated fatty acid oxidation. In general, proteins are considered tasteless at secondary, tertiary, and quaternary structures. The bitter peptides are normally not exposed, thus they yield no bitterness. During further processing, specific application conditions may trigger the release of bitter peptides. Microorganisms, proteases, peptidases, and acids can hydrolyze protein to free some bitter peptides that contribute to an unpleasant bitter taste.

Fujimaki et al.[14] identified the bitter peptides in soybeans as H·Gly–Leu·OH, H·Leu–Phe·OH, H·Gly–Leu–Lys–Ser·OH, H·Leu–Lys·OH, H·Gln–Gly–Ile-2·OH, H·Leu–Phe–Val·OH, H·Arg-2–Leu·OH, and H·Arg–Leu·OH. Free amino acids, such as isoleucine, leucine, phenylalanine, and valine, have also been identified as contributing to bitterness. Most of the bitter peptides were found to have leucine at either terminal. All of these peptides had molecular weights under 1500 and were difficult to further hydrolyze by endopeptidase.

Hydrophobic interactions have been identified as contributing factors to protein behaviors. The hydrophobic residues of amino acids in a peptide are driven together by clusters of water molecules so as to form secondary peptide or protein structures. Hydrophobicity can also be used as a predictor of bitterness of a protein, peptide, or amino acid. The average hydrophobicity of a peptide, or Q value, was introduced by Ney[15] to evaluate the prediction accuracy for various proteins. The Q value is obtained by dividing the sum of the f values of a single amino acid by the number of amino acid residues, as shown in Equation 12.1. Here, the f value represents a measure of the hydrophobicity of an amino acid residue:

$$Q = (\Sigma \Delta f)/n \qquad (12.1)$$

The Q values of bitter peptides were much higher than those for non-bitter peptides. The Q value was used to predict peptide bitterness in various proteins, such as casein, K-casein, spinach, soy, potato, wheat, and zein; however, the position of the amino acid and molecular weight were found to have no influence on bitterness.[15]

Research has shown that all bitter peptides have Q values greater than 1400, but no peptide with a Q value less than 1300 is bitter. For Q values between 1300 and 1400, no prediction can be made regarding bitterness. This rule is also applicable to individual L-amino acids for which the Q value equals the f value, with the exceptions of lysine and praline. Their Q values are too high not to be bitter, and both have a sweetish taste with slight bitterness. It is possible that the sweetness overcomes the bitterness. Q values also can be applied to proteins when also taking into consideration molecular weight to predict the possibility of producing bitter hydrolysates.[15]

Lipoxygenase and peroxydase can catalyze linolenic acid to produce bitter trihydroxyoctadecenoic acid. The increased hydroxyl group induces bitterness of the lipid compounds.[15] Because the hydrophobicity of lipid base compounds is too great to use Q values, R values (the ratio of total carbon atoms to total hydroxyl groups) can be used to predict the bitterness of lipid compounds, as shown in Equation 12.2:

$$Q = n_c/n_{OH} \qquad (12.2)$$

For phosphatides, one phosphatidylcholine is equivalent to two hydroxyl groups. When the Q value exceeds 7, no bitterness is detected.

Almost all the bitter compounds, both protein and lipid, can be controlled. Reducing the bitterness of soy products is as important as doing so with other off-flavors. During processing, specific procedures can be implemented to reduce lipoxygenase activity or prohibit the release of bitter peptides. This is discussed further in the section on processing treatments to reduce off-flavors.

Soy Protein Product Flavor-Binding Capacity

Soy protein products are widely used as functional ingredients in matrices to stabilize the system or improve nutritional values. The effects of soy protein on food flavors are as important as its functional properties in food matrices. Detailed investigations into the interactions between soy protein products and flavors are essential. Furthermore, studying the effects of other basic ingredients on the system is critical to understanding the complexity of the interactions. All food ingredients — water, proteins, lipids, carbohydrates, sugars, and salts — interfere with flavor compounds. Each individual component has a different response to flavor compounds depending on the physicochemical properties of the food components and flavors.

The addition of inorganic salts can increase the vapor pressure of the volatile compounds in dilute water solutions; however, the effect of sugars in a diluted aqueous system is more complex. They may increase or decrease vapor pressure, which is compound dependent. Nawar[16] observed that the vapor concentration of methyl ketones in water solution is greater with increasing molecular weight. This study also observed that additional sucrose increases acetone headspace concentrations but depresses heptanone and heptanal headspace concentrations.

Headspace analysis of food volatiles is popular in flavor research. The headspace concentration is proportional to the pure solute concentration in water. Volatile compounds in diluted solutions have similar relationships. These relationships can hold true in food systems, as well. A change in headspace concentration can reflect the flavor concentration in food; however, the relationship between volatile compounds and water is more complicated in food than in pure solution due to the coeffects of other components, such as proteins, carbohydrates, lipids, salts, sugars, and other functional binding ingredients. The complexity of food systems is a result of the multiple interactions of various compounds.

Controlling the interactions among food compounds is critical to achieving flavor balance in food. Studies show that casein, gelatin, ovalbumin, carbohydrates, α-lactalbumin, bovine serum albumin, alfalfa leaf protein, single-cell protein or yeast, and soy proteins are all involved in headspace flavor concentration.[17] Not only the ingredients but also their order of addition can have a significant impact upon the volatility of the flavors in foods.[16,17] The proteins α-lactalbumin, bovine serum albumin, alfalfa leaf protein, single-cell protein or yeast, and soy proteins consistently decrease the headspace concentrations of flavors (hexanal, heptanal, octanal, 2-hexanone, 2-heptanone, and 2-octanone) in model water and flavor systems.[17] Proteins reduce the dispersion of flavors in water by binding or trapping flavors in their structures. Sources of lipids in food systems can also affect flavor release or binding. Endogenous lipids in alfalfa leaf protein concentrate have been shown to enhance interactions between proteins and flavors; however, the addition of proteins to the flavor and oil mixture did not change the volatility of the flavors.[17] A similar phenomenon was observed by Nawar[16] when heptanone was added into a sucrose solution. The headspace concentration was considerably higher than when sucrose was added into a heptanone solution. Pentanone caused a similar response, while acetone and butanone behaved differently — regardless of the order of addition, the headspace concentration increased, and the order of addition of the compounds had no effect on the solution volume. After 24 hours of equilibration, the headspace concentration remained the same, suggesting possible binding effects between the volatiles and the sugars. Without water, the sugar effects disappeared. Such an influence on headspace may be due to water and ingredient interactions such as competitive repulsion and attraction. Gelatin and glycerol were found to have less of an impact on the volatile headspace concentration.

The presence of gelatin or soy protein decreases the volatility of methyl ketones,[16] and soy protein decreases aldehyde volatility.[18] It can be said that, as the chain length of the aldehyde increases, interactions between the soy protein and aldehydes also increase, and the aldehydes are less volatile. The headspace flavor concentration decreases with the addition of soy protein.[17] Quantitative flavor–protein interaction studies show that soy protein has four binding sites for all the carbonyls.[19] The binding constants increase with the chain length of the ligand by three orders of magnitude for each methylene group increase in the chain. The favorable change in the hydrophobic free energy is 550 cal/mol of each CH_2 residual. The position of the carbonyl group decreases the hydrophobic free energy by 105 cal/mol of each carbon shifted away from the terminal 1 end. Binding affinity is independent of temperature above 25°C, and a drastic increase in the binding affinity is observed at 5°C. Partially denatured soy protein increases the binding constant. Thermodynamic analysis of the binding of carbonyls with soy protein has shown it to be relatively weak. Soy protein isolates and textured soy protein have similar binding patterns and strength, but soy concentrates bind more flavors than they do. The 70% protein and 20% carbohydrate content of the soy protein concentrate might explain the higher flavor-binding capacity.[17]

The thermodynamic study of soy protein has shown no significant differences in homologous heat adsorptions of aldehydes, ketones, and methyl esters; however, the alcohols adsorbed the greatest and hydrocarbons the least. The functional group of the ligand plays a significant role in binding flavor compounds to soy protein in the dry state.[19] The binding affinity of the ligand for a protein is dependent on the structural state of the binding sites. Any change in the native conformation of the protein could affect the binding affinity. Most of the commercially prepared soy protein products are partially denatured. Partially denatured soy protein isolates have a higher binding affinity than do native soy proteins.[20] Arai et al.[21] reported that heat denaturation increases the resistance to removal of off-flavors by vacuum distillation. Damodaran and Kinsella[20] suggested that the number of binding sites for 2-nonamone in soy protein does not change, whereas the binding affinity of the ligand to the same sites increases by 30%. This is possibly due to the reorganization of subunits that enhances the hydrophobicity of the existing hydrophobic sites and thus increases the binding affinity for the ligands.

Soy protein does not bind alcohols, binds aldehydes both reversibly and irreversibly, and binds ketones reversibly.[18] The headspace concentration is reduced in the presence of protein.[22] Research[23,24] suggests that soy protein adsorption is nonspecific (plus one hydrogen bond for carbonyls or two hydrogen bonds for alcohols); furthermore, soy protein has numerous polar binding sites that facilitate the formation of two hydrogen bonds.

The retention of flavor compounds *n*-hexanal and *n*-hexanol in soy products increases, in order, from native soy protein to partially denatured soy protein to denatured soy protein. Arai et al.[21] found that a vacuum enhances retention;

however, Crowther et al.[24] showed that heated proteins have fewer adsorption coefficients for hexanal and thus less binding. Arai et al.[21] detected chemical bonding, while Crowther et al.[24] measured reversible adsorption. Regardless, it can be said that processing conditions are critical to flavor compound retention in soy protein. At the same time, chemical and physical properties of the flavor compounds will trigger binding just as do physicochemical properties of soy protein products. Hexanal is more reactive than hexanol, but hexanol is adsorbed at a greater rate than hexanal.[24] Heat-treated soy protein displays more differences in physicochemical properties in comparison with untreated soy proteins. It can be said that less protein preprocessing results in a more drastic protein response during further processing.

Identifying soy protein process conditions and the chemical compounds involved in the process is the first step in determining the possible off-flavors that soy proteins could bring into the food product. Second, the potential soy protein flavor-binding capacity has to be considered in balancing the overall flavors of finished products. Because protein concentration, degree of denaturation, structure configuration, and so on all affect the physical, chemical, and physicochemical properties of soy protein, soy protein products can respond differently to various flavor compounds. Being aware of all the off-flavors that could possibly be contributed by soy protein to the food system is a critical step toward reducing off-flavors. Furthermore, food systems are quite complex because each component makes a unique contribution to the overall food characteristics. It is possible to predict possible antagonistic and synergistic interactions between both volatile and nonvolatile flavor compounds and food systems with complex proteins, carbohydrates, lipids, and water matrices. By understanding such responses, correct balancing of the overall flavors can be achieved.

Information gained through research can be critical to designing soybean processing procedures to prevent oxidation and eliminate the generation of off-flavor compounds. Processing manipulation can minimize lipoxygenase activities, thereby eliminating off-flavor generation and enhancing soy product functionality. When substrates are available, enzyme activities depend not only on pH but also on temperature and time allowed for reactions. Restricting oxygen, adjusting the pH, and controlling the temperature and processing dwell time are all significant control factors to be considered when trying to prevent lipid oxidation.

Processing Treatment To Reduce Off-Flavors

Because most off-flavors are formed during the initial mechanical processing of soybeans, researchers have investigated various methods to control lipoxygenase activities and prevent off-flavor generation. High-temperature and acid treatments are commonly used to denature lipoxygenase. During heat treatment, dwell time is critical to obtaining optimum results with the least protein denaturation. Acid treatment effectiveness is dependent on the type of acids used. A disadvantage of these treatments is the protein denaturation, which lowers both the nutritional value and processing functional properties. Acid treatment could also increase the total salt content of soy protein products. Without early treatment, most off-flavor compounds are removed during oil extraction; however, various substances may be bound or entrapped in protein and carbohydrate matrices, especially for wet milling. Due to the low threshold and high impact of some compounds on soy protein flavors, further treatments are often necessary to disassociate and remove certain substances.

Organic solvent extraction, supercritical CO_2, and enzyme treatments are among corrective treatments for disassociating and removing off-flavor compounds from soy protein products.[8,14,21,25,26] Removing these flavors, especially off-flavor compounds remaining in the soy protein, is important in improving soy protein product qualities. All of these treatments possibly improve the overall flavors of soy protein products; however, each treatment can have a different impact on the product properties. Research has demonstrated the various impacts on functional properties, flavors, and taste. Different processing treatments are discussed here in an effort to evaluate the merits of each treatment, understand the different effects on soy products, and determine specific treatments for individual soy protein products.

Processing Temperature Control

Major off-flavors are generated by lipoxygenases following direct contact with unsaturated fatty acids during soybean processing. Deactivating lipoxygenase is one effective method for preventing off-flavor compound generation. Processing conditions greatly affect lipoxygenase activities; therefore, adjusting and controlling processing parameters can be the simplest methods for minimizing off-flavors. Both dry and wet heat treatments can inactivate lipoxygenases. Such treatment may reduce the solubility of soy proteins and also create cooked and toasted flavors that are not desirable. Thermal inactivation of lipoxygenase also maintains protein quality.

Traditionally, soymilk is processed by soaking the soybeans in water for hours followed by vigorous grinding, filtering, and heating to a boil within a room temperature environment. During grinding, extensive cell-wall breakage causes lipoxygenases to catalyze lipid hydroperoxidation and expose unsaturated fatty acids to autoxidation. In the absence of pretreatment, they rapidly develop undesirable odors and off-flavor which persist even after cooking. Mizutani and Hashimoto[5] investigated the effect of grinding temperature on off-flavors during soymilk manufacturing and found that hydroperoxidation correlates with off-flavor content. At 3°C and 80°C, the amount of hydroperoxides was about half that at 30°C; *n*-hexanal showed the highest correlation with hydroperoxides ($r = 0.96$), except at 80°C. There are two possible reasons for the high-temperature reduction of *n*-hexanal.

First, *n*-hexanal evaporates at high temperature; second, heat-denatured protein has a much greater capability to bind off-flavors. Heat-denatured soy protein binds *n*-hexanal at a rate about three times greater than that of native proteins, while heat-denatured soy protein with ethanol combines with *n*-hexanal at a rate about six times greater than that of native protein.[21] Both high- and low-temperature grinding effectively controls hydroperoxidation and off-flavors; however, quantitative evaluations of the soluble protein nitrogen content showed that samples ground at 80°C had only 42.7% the content at 3°C. Protein heat denaturation can cause reduced protein solubility;[5] nevertheless, protein denaturing before protein solubilization reduces nitrogen extraction and gel formation ability. Natural soybean hydroperoxidation increases with temperature and reaches a maximum at 50°C, then decreases sharply with further temperature rise. L-2 activity reaches a maximum at about 30°C.[9] Research has shown that grinding soybeans at a low temperature, such as 3°C, maintains protein solubility and gel formation ability and can produce soymilk with a high protein concentration and low off-flavor.[5,7]

Studies of the stability of whole soy milling of full-fat soy flour showed that individual dry or wet heat treatment can produce soy flour with a low peroxide and free fatty acid content, good shelf stability, and acceptable flavor.[27] The peroxide content of whole soybeans increases with water soaking time; additional wet heat treatment after dry heat treatment results in poor flavor stability.

The solubility of soy protein depends on the type of milling process. If denaturizing occurs before the soy protein is dissolved in water, the product solubility most likely will be lowered. Solubility also depends on the availability of water during processing. Protein solubility, gel forming ability, and biovalue are all important factors in evaluating soymilk qualities; therefore, grinding soybeans at low temperatures can provide soymilk with high protein concentration, good gel-formation ability, and less off-flavor. Low-temperature grinding can also provide better soymilk for tofu and other soy foods made from soymilk. Low-temperature grinding, such as at 3°C, is an economical processing method for producing soymilk with good protein solubility and gel-formation ability, a high protein concentration, and few off-flavors.

Acid Treatment: pH Adjustment

Most soybeans are processed at a pH of 6 to 7, and L-2 and L-3 are more responsible for off-flavor generation during processing at pHs in this range. Adjusting the pH can be a key method for reducing oxidative off-flavors during grinding. Kon et al.[28] observed that adjusting the pH to 3.85 or below significantly controlled off-flavor production during raw soybean grinding. No off-flavors were detected in soybean milk and slurry acidified to below 3.85, according to both sensory and gas chromatography (GC) evaluations. A lack of pentanal and hexanal peaks, corresponding to acetaldehyde, in acid-treated samples suggests that the rancid and painty odorants generated during the oxidative deterioration of lipids were eliminated. The correlation between sensory and GC results also suggests that hexanal is more responsible for soy protein off-flavors. At pH 3.85, soy protein extraction was found to be poor, directly affecting production yield and qualities of soy protein products. Protein extraction and solubility are important characteristics of soy protein products and are related to nutritional value and ingredient functionality. Soy protein extraction is lowest at the isoelectric point, a pH of approximately 4 to 4.5. Avoiding processing soy protein at a pH close to the isoelectric point is a must for maintaining soy protein extraction. Kon et al.[28] observed that the maximum protein extraction was achieved at pH 2.1. Adjusting the pH can be an economical way to achieve better quality soy protein products with fewer off-flavors, and the type of acids used to adjust the pH affects the formation of off-flavor compounds. Kon et al.[28] concluded that HCl performed the best. Acid treatment may dramatically reduce soy protein dispersibility, depending on the pH and chemical reagent used for neutralization. Che Man et al.[29] observed irreversible inactivation of lipoxygenase at pH 3.0 or below. After neutralization, lipoxygenase activity will recover to a certain extent with treatment above pH 3, and in this study soybeans treated with various acids (hydrochloric, phosphoric, and tartaric) demonstrated similar denaturing patterns. Denaturation occurred rapidly with early pH drop. The protein dispersibility index (PDI) showed that acid-treated soybean neutralized with potassium (69.3%) was more soluble when followed by sodium (68.6%) and calcium (62.5%). Using potassium and calcium can reduce sodium involvement, thus addressing a health concern. Dialyzed samples showed an increased PDI of 7 to 8%; however, dialysis is very practical for industry use.

Supercritical Carbon Dioxide Extraction

Supercritical fluid treatments are becoming popular in the food industry for various reasons. Using inorganic chemical compounds under particular physical conditions to form supercritical fluids allows compounds to have the same properties as organic solvents; therefore, certain strongly bound chemical compounds can be disassociated and removed from the carrier. Supercritical carbon dioxide ($SC-CO_2$) can be used to remove off-flavors from flour, corn meal, and soy proteins. The physical conditions of the supercritical fluid (pressure, temperature, and density) and its ratio to soy protein are critical to the quality of soy protein products. High-temperature and high-pressure $SC-CO_2$ improved the overall flavor profile of corn products; however, temperatures over 80°C and pressures over 64 MPa dramatically increased protein denaturation, leading to a considerable decline in the processing functional performance.[30] Maheshwari et al.[8] investigated the use of $SC-CO_2$, liquid CO_2 ($L-CO_2$), and $SC-CO_2$ with ethanol ($SC-CO_2r/EtOH$) under conditions of lower pressure and temperature to remove off-flavors of soy protein isolate and concluded that SC-CO_2 was most effective and had no detrimental effects on the soy protein isolate processing functionality. The conditions used in this study were 17.2 MPa and 25°C for L-CO_2 and 27.6MPa and 40°C for both SC-CO_2 and SC-CO_2/EtOH. Sensory analysis showed improvement of treated samples with increased overall acceptability. The significant reduction in moisture content of samples treated by SC-CO_2 or SC-CO_2/EtOH suggested that adjusting soy protein isolate moisture content prior to supercritical fluid deodorant treatment was necessary to maintain the standard moisture content of the finished products. A reduced moisture content can mean higher costs and lower profits. A significant drop in pH in all treated samples from 7.03 to between 6.70 and 6.86 was observed, possibly due to CO_2 absorption of the soy protein.

Headspace gas chromatography data have confirmed sensory evaluations with regard to the significant reductions in off-flavor compounds. Among the removed compounds are medium-chain aldehydes, ketones, and alcohols. Medium-chain aldehydes, *n*-butanal, *n*-pentanal, and *n*-hexanal are the major contributors to off-flavors. L-CO_2 was least effective in all cases. The removal of total identified volatiles reached a maximum (72% of control sample) through SC-CO_2 treatment at a high gas and protein ratio. SC-CO_2 and SC-CO_2/EtOH treatments effectively removed pentanal and hexanal, while EtOH further enhanced effectiveness. However, the EtOH residual could carry alcohol flavor to the products. Data have shown that both SC-CO_2 and SC-CO_2/EtOH treatment efficiently disassociate the ketones. SC-CO_2 treatment was found to remove 2-butanone, 2-pentanone, and 2-hexanone by 94%, 86%, and 71%, respectively. Alcohol removal was the most difficult using these methods. Under the highest volume of SC-CO_2, only 72% of 1-butanol and 56% of 1-pentanol were removed. The high polarity of butanol and pentanol may account for the difficulties. It can also be said that alcohol binds soy protein more strongly than aldehydes and ketones do. Ethanol is also less supportive in improving the SC-CO_2 treatment for alcohol compound removal. SC-CO_2 is most effective in removing aldehydes and less effective in removing alcohols. This is consistent with the binding constants of these compounds with soy proteins. The binding constants of off-flavor ligands were found to decrease, in order, with aldehydes, ketones, and alcohols. The sensory data agree with the analytical data in that high-level SC-CO_2 treatment yielded the most acceptable products. Because water is a critical ingredient in triggering changes in food compounds, further evaluation of supercritical-fluid-treated soy protein isolates in food processing is needed to establish the advantage of such treatment.

Enzyme Treatment

The capacity of soy protein to bind to certain substances can be strong enough to resist simple physical treatments, such as distillation and extraction. Bound and trapped flavors may be disassociated partially during physical treatment or possibly released during further food processing and consumption when other supporting factors are introduced. Various lipid substances can bind to soy proteins during early processing to cause off-flavors or generate more off-flavor compounds when conditions are suitable. Most phospholipids in soybeans are difficult to extract with hexane or other similar nonpolar solvents used for oil extraction. These unsaturated fatty acids of phospholipids can act as precursors to thiobarbituric acid (TBA)-positive substances. Enzymes can disassociate these bound compounds from soy proteins. Keeping in mind the ability of soy protein to bind lipids and flavor compounds, enzymatic treatment is an economical method for improving the flavor quality of soy protein products. Researchers have investigated the effectiveness of different enzymes in liberating odorants and reducing off-notes. Extensive research has been conducted into enzyme treatments. Aspergillopeptidase,[21,31] molsin,[21] porcine liver aldehyde oxidase,[32] and aldehyde dehydrogenase[33] were found to effectively reduce the off-flavor compounds.

Noguchi et al.[31] investigated the use of aspergillopeptidase A (APase A) to partially digest tofu and soybean flour to remove flavor compounds and related fatty materials. The amounts of ether-soluble compounds, carbonyl, and volatile-reducing substances released from both soybean products were greater than for the control sample during incubation. Sensory evaluation showed that treated samples had fewer off-flavors, less taste, and less color. Lipids, including triglycerides, diglycerides, fatty acids, phosphatidylcholine, phosphatidylethanolamine, and sitosteryl glucoside, were released from treated samples as well; therefore, further lipid oxidation can also be eliminated. The ratio of enzymes to substrate as well as incubation time are critical to the effectiveness of enzyme treatments. In general, 1% enzymes and up to 2 hours' incubation time were sufficient. With prolonged incubation over 4 hours, the sample began to develop bitterness, and after 8 hours a maltol-like flavor developed. Among released compounds, saponin pigments were also detected. Both sensory and TBA values indicate that soy products treated with APase A are more stable and have fewer off-flavors.

Similar results were obtained by Aria et al.[21] APase A was found effective in removing odors such as *n*-hexanal, *n*-hexanol, and *n*-heptanol from soy protein isolates. They also investigated differences between molsin, a crude prepared APase A, and crystallized APase A. The molsin-treated sample was less bitter and contained greater amounts of free amino acids, especially hydrophobic alanine, valine, isoleucine, leucine, tyrosine, and phenylalanine. Soy protein treated with crystallized APase A was much more bitter and had lesser amounts of free amino acids, mainly hydrophilic aspartic acid, glutamic acid, asparagines, glutamine, serine, and threonine. Soy protein isolates treated with APase A had high levels of peptides and bore hydrophobic amino acid residues near the C-terminal. The hydrophilic-to-hydrophobic free amino acid ratio of APase A-treated soy protein was 3 to 1, and the ratio for molsin-treated soy protein was 3 to 4. Molsin contains APase A and carboxypeptidase, identical to aspergillus acid carboxypeptidase (AACPase). AACPase is an exopeptidase that disassociates amino acids one by one from the C-terminal of the peptide. The reduction in bitter peptide concentration lessened the bitterness. Although, as mentioned, soy protein can have bitter amino acids and peptides, molsin not only removed these off-flavor compounds from soy proteins but also disassociated the bitter peptide structure. It can be said that a combination of enzyme treatments of soy proteins can create more desirable and acceptable products.

Maheshwari[32] investigated using porcine liver aldehyde oxidases (PAOs) to reduce the off-flavors of soy proteins and discovered that PAO-I was more substrate specific than PAO-II. PAO-I is more stable at pH 9 than 7. Adjustment of pH is often necessary to achieve maximum reaction results. It was found that 0.5% PAO-I significantly reduced off-flavor-causing aldehydes (acetaldehyde, n-propanal, n-butanal, n-pentanal, and n-heptanal) at pH 9 with 60 minutes' incubation. The response of each aldehyde was also time dependent, so controlling incubation time is very important. As a result of this treatment, carboxylic acid was detected in treated soy products. The higher energy of activation for the conversion of aldehydes to carboxylic acids (47 kJ/mol) by PAO-I compared with the peroxidation of linoleic acid (18 kJ/mol) by lipoxygenase suggested that off-flavor removal occurs faster than off-flavor generation. The sensory data also showed significant improvement and a great beany note reduction in soybean protein treated with PAO-I. Enzyme treatments can be introduced to certain soy protein products, especially when the free amino acids and peptides are complementary in terms of taste and nutritional values in finished products.

Genetic Mutants

Biotechnology has produced new soybean varieties with significantly reduced content stachyose and raffinose (sugar) content, resulting in major flavor improvements. When bioengineered soybeans are used for manufacturing functional soy protein ingredients, improved taste and better digestibility allow higher inclusion levels when desired. Genetically modified soybeans free of L-2 showed the least accumulation of n-hexanal; those that were free of L-1 and L-2 had a moderate but significant role in hexanal production. The lack of L-2 and L-3 did not affect hexanal production or TBA value. In other words, L-2 is most responsible for hexanal generation.[34] Mutant soybeans lacking lipoxygenases are a good means of preventing off-flavor in soy products. No physiological or agronomical consequences of soy beans being genetically modified to eliminate individual lipoxygenase were detected;[34] however, further studies are necessary to determine the impact of mutations on soy crop yield, nutrition, quality of soy products, and other functional aspects. Concerns regarding genetic modification include public acceptance and unknown consequences.

Masking and Creating Flavors

The nature of food can be very complex. The molecular weight, structure, hydrophobicity, and hydrophilicity of flavors and food macro compounds;

food ingredient interactions; and processing parameters are all critical factors in overall food flavor and flavor stability. Determining the antagonistic and synergistic interactions among major food components and flavors can lead to ways to eliminate or reduce the contribution of soy protein to overall flavors. Deactivation of lipoxygenase can only prevent internal off-flavor generation. Processing treatments can reduce off-flavors dramatically; however, during transportation and further processing, soy protein products still have the ability to absorb and entrap odorants from packaging materials and their surroundings, which has a negative impact on food application. Further preventive treatments are critical to maintain, stabilize, and enhance food products with soy as key ingredient.

Masking Bitterness

Bitter taste depends on the ammonium groups of peptides and amino acids. The transition from amino acids to corresponding amines was shown to yield a decrease in bitterness.[35] The bitter-flavor masking activity of glutamic-acid-related compounds has been reported by researchers.[15,36] The masking action of peptides rich in glutamyl residues was observed in plastein synthesis, an endopeptidase-catalyzed reverse reaction;[36] however, plasteins introduced bitterness when further hydrolyzed.[21] Noguchi et al.[36] introduced a glutamic-acid-rich oligopeptide fraction into a fish protein system and reconstituted plasteins. When the glutamic-acid-rich plastein was hydrolyzed again, it had no bitterness. A glutamic-acid-rich oligopeptide, L-glutamic acid diethylester, altered the acidic, basic, and neutral fraction ratios. With increased acidic fraction and low molecular weight, no bitterness was detected, but the product had a slightly brothy taste, and it was free of nonbinding glutamic acid. Sensory evaluations showed that these acidic fractions had a masking function similar to that of L-glutamyl-L-glutamic acid against various bitter compounds including L-isoleucine, Gly–Leu, magnesium chloride, chlorogenic acid, caffeine, phenylthiocarbamide, and brucine. Such a method can be used in soy protein processing to prevent bitterness arising during processing and especially in savory products, where a brothy taste can have a very positive impact. The slight broth taste can only enhance the overall savory impact of the products.

Paths to Creating Soy Product Flavors

In general, sugar, salt, sauces, condiments, seasonings, vegetable and fruit purees and juices, other foods, or food components are used to cover up the beany, grassy, stale, earthy, bitter taste of soybean and soy products; however, such covering methods failed to truly resolve the off-flavor problem. As discussed earlier, both volatile and nonvolatile compounds are responsible for off-flavors, and further processing of food products incorporating soy products still faces challenges in preventing the generation of new off-flavors and achieving a balance of flavors with high consumer acceptance. Following processing treatment, soy products require further processing for stabilization. Hydrogenation can stabilize soy lecithin. A substance with a high binding affinity to soy protein and positive flavor and taste impact is needed to stabilize the soy protein without altering its functional performance. Because soy proteins interact greatly and influence flavor volatility and other physicochemical properties, the first step is to define specific applications of soy protein products in food formulas. When the applications are defined, flavor compounds can be used to treat soy protein products to create suitably flavored ingredients having the desired food flavors. Such ingredients would provide functional support without altering the original food flavor or having a negative impact; for example, if soy protein isolates had a taste similar to that of dehydrolyzed nonfat milk powder, it could be used as a nonfat milk powder replacement without too much flavor reformulation.

Soy protein is often incorporated to enhance the stability of food products, but soy ingredients have a significant effect on flavors that directly affects overall product flavor profiles. Practical solutions involve early processing treatment as well as further processing during the manufacture of the food product. Adequate reduction or masking of off-flavors through processing without changing the functionalities of the soy protein products will improve the product. An ideal masking agent must overcome or eliminate off-flavors to create high-quality products. A unique flavor system would have optimal flavor impact and would interfere minimally with other ingredients. On the other hand, discovering new soy protein product properties and turning their drawbacks into advantages could yield stable foods with desirable texture and flavor. Given the market demand for soy products as functional ingredients, greater systematic manipulation, protection, and modification during food processing are required.

References

1. Brewster, E., Soy consumers count on good taste, in *Stagnito's New Products Magazine Soy Supplement*, Stagnito Communications, Deerfield, IL, 2004, p. s8.
2. Suriyaphan, O., Drake, M.A., and Cadwallader, K.R., Identification of volatile off-flavors in reduced-fat cheddar cheeses containing lecithin, *Lebensm.-Wiss. U.-Technol.*, 32, 250, 1999.
3. Messina, M., *Soy Protein and Cholesterol Reduction*, Archer Daniels Midland Company (ADM), Decatur, IL, 2003.
4. Perkins, E., Composition of soybeans and soybean products, in *Practical Handbook of Soybean Processing and Utilization*, Erickson, E.R., Ed., AOCS Press, Champaign, IL, 1995, chap. 2.

5. Mizutani, T. and Hashimoto, H., Effect of grinding temperature on hydroperoxide and off-flavor contents during soymilk manufacturing process, *J. Food Sci.*, 69, SNQ112, 2004.

6. Matoba, T., Hidaka, H., Narita, H., Kitamura, K., Kaizuma, N., and Makota, K., Lipoxygenase-2 isozyme is responsible for generation of *n*-hexanal in soybean homogenate, *J. Agric. Food Chem.*, 33, 852, 1985.

7. Samoto, M., Miyazaki, C., Kanamori, J., Akasaka, T., and Kawamura, Y., Improvement of the off-flavor of soy protein isolate by removing oil-body associated proteins and polar lipids, *Biosci. Biotechnol. Biochem.*, 62, 935, 1998.

8. Maheshwari. P., Ooi, E.T., and Nikolov, Z.L., Off-flavor removal from soy-protein isolate by using liquid and supercritical carbon dioxide, *J. AOCS*, 72, 1107, 1995.

9. Furuta, S., Nishiba, Y., Hajika, M., Igita, K., and Suda, I., DETBA value and hexanal production with the combination of unsaturated fatty acid and extracts prepared from soybean seeds lacking two or three lipoxygenase isozymes, *J. Agric. Food Chem.*, 44, 236, 1996.

10. Axelrod, B., Cheesbrough, T.M., and Laakso S., Lipoxygenase from soybeans, *Methods Enzymol.*, 71, 441, 1981.

11. Nash, A.M., Eldridge, A.C., and Wolf, W.J., Fractionation and characterization of alcohol extractables associated with soybean proteins: nonprotein components, *J. Agric. Food Chem.*, 15, 103, 1967.

12. Anderson, R.L. and Warner, K., Acid-sensitive soy protein affect flavor, *J. Food Sci.*, 41, 293, 1976.

13. Sucan, M.K., Identifying and preventing off-flavors, *J. Food Technol.*, 58, 36, 2004.

14. Fujimaki, M., Yamashita, M., Okazawa, Y., and Arai, S., Applying proteinolytic enzymes on soybean. 3. Diffusible bitter peptides and free amino acids in peptic hydrolyzate of soybean protein, *J. Food Sci.*, 35, 215, 1970.

15. Ney, K.H., Bitterness of peptides: amino acid composition and chain length, in *ACS Symposium Series*, Boudreau, J.C., Ed., American Chemical Society, Washington, D.C., 1979, chap. 6.

16. Nawar, W.W., Some variables affecting composition of headspace aroma, *J. Agric. Food Chem.*, 19, 1057, 1971.

17. Franzen, K.L. and Kinsella, J.E., Parameters affecting the binding of volatile flavor compounds in model food system. I. Protein, *J. Agric. Food Chem.*, 22, 675, 1974.

18. Gremli, H.A., Interaction of flavor compounds with soy protein, *J. AOCS*, 51, 95a, 1974.

19. Damodaran, S. and Kinsella, J.E., Interaction of carbonyls with soy protein: thermodynamic effects, *J. Agric. Food Chem.*, 29, 1249, 1981.

20. Aspelund, T.G. and Lester, A.W., Adsorption of off-flavor compounds onto soy protein: a thermodynamic study, *J. Agric. Food Chem.*, 31, 539, 1983.

21. Arai, S., Noguchi, M., Kurosawa, S., Kato, H., and Fujimaki, M., Applying proteolytic enzymes on soybean. 6. Deodorization effect of aspergillopeptidase A and debittering effect of aspergillus acid carboxypeptidase, *J. Food Sci.*, 35, 392, 1970.

22. Franzen, K.L. and Kinsella, J.E., Physicochemical aspects of food flavoring, *Chem. Ind.*, 21, 505, 1975.

23. McMullin, S.L., Bernhard, R.A., and Nickerson, T.A., Heats of adsorption of small molecules on lactose, *J. Agric. Food Chem.*, 23, 452, 1975.

24. Crowther, A., Wilson, L., and Glatz, C., Effects of processing on adsorption of off-flavors onto soy protein, *J. Food Proc. Eng.*, 4, 115, 1980.
25. Smith, A.K. and Circle, S.J., Protein products of food ingredients, in *Soybeans: Chemistry and Technology*. Vol. 1. *Proteins*, Smith, A.K. and Circle, S.J., Eds., AVI Publishing, Westport, CT, 1978, p. 339.
26. Eldridge, A.C., Kalbrenner, J.E., Moser, H.A., Honig, D., Pakis, J.J., and Wolf, W.J., Preparation and evaluation of supercritical carbon dioxide defatted flakes, *J. Food Sci.*, 51, 584, 1986.
27. Mustakas, G.E., Alberetch, W.J., McGhee, J.E., Black., G.N., Bookwalter, G.N., and Griffin, Jr., E.L., Lipoxidase deactivation to improve stability, odor, and flavor of full fat soy flour, *J. AOCS*, 46, 623, 1969.
28. Kon, S., Wagner, J.R., Guadagni, D.G., and Horvat, R.J., pH adjustment control of oxidative off-flavors during grinding of raw legume seeds, *J. Food Sci.*, 35, 343, 1970.
29. Che Man, Y.B., Wei, L.S., and Nelson, A.I., Acid inactivation of soybean lipoxygenase with retention of protein solubility, *J. Food Sci.*, 54, 963, 1989.
30. Wu, Y.V., Friedrich, J.P., and Warner, K., Evaluation of corn distillers' dried grains defatted with supercritical carbon dioxide, *Cereal Chem.*, 67, 585, 1990.
31. Noguchi, M., Arai, S., Kato, H., and Fujimaki, M., Applying proteolytic enzymes on soybean. 2. Effect of aspergillopeptidase A preparation on removal of flavor from soybean products, *J. Food Sci.*, 35, 211, 1970.
32. Maheshwari, P., Murphy, P.A., and Nikolov, Z.L., Characterization and application of porcine liver aldehyde oxidase in the off-flavor reduction of the soy proteins, *J. Agric. Food Chem.*, 45, 2488, 1997.
33. Chiba, H., Takahashi, N., and Sasaki, R., Enzymatic improvement of food flavor. II. Remove beany flavor from soybean products by aldehyde dehydrogenase, *Agric. Biol. Chem.*, 43, 1883, 1979.
34. Moreira, M.A., Tavares, S.D., Ramos, V., and Barros, E.G., Hexanal production and TBA number are reduced in soybean [*Glycine max* (L.) Merr.] seeds lacking lipoxygenase isozymes 2 and 3, *J. Agric. Food Chem.*, 41, 103, 1993.
35. Belitz, H.-D., Chen, W., Jugel, H., Treleano, R., and Wieser, H., Sweet and bitter compounds: structure and taste relationship, in *ACS Symposium Series*, Boudreau, J.C., Ed., American Chemical Society, Washington, D.C., 1979, chap. 4.
36. Noguchi, M., Yamashita, M., Arai, S., and Fujimaki, M., On the bitter-masking activity of a glutamic acid-rich oligopeptide fraction, *J. Food Sci.*, 40, 367, 1975.

13

Selecting Soybeans for Food Applications

Lynn Clarkson

CONTENTS

Introduction

When the product, process, and client have been determined, the immediate challenge becomes one of optimizing the product and production process while distinguishing the product from those of others and cementing a good relationship with the client. Raw material selection addresses each aspect of this challenge (Figure 13.1). Sophisticated buyers select soybeans by variety and production geography. They spurn commodity soybeans as inadequate

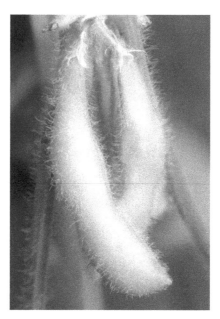

FIGURE 13.1
Healthy soybeans growing in the pod on their way to becoming tofu. This variety is preferred by several Japanese companies to make tofu and is being produced under contract for a Japanese processor.

and established grade standards as irrelevant. They use contract production and identity-preserving supply systems to guarantee availability of their preferred soybeans when they want them. Such buyers have done so for centuries to obtain better flavor, nutrition, consistency, process yield, market access, market differentiation, and improved income. More recently, food buyers have begun to use the same systems to ensure traceability as well as compliance with cultural standards such as organic certification and the avoidance of certain pesticides or genetically engineered traits. To control raw material issues, buyers develop a supply chain that starts with a contract manager and includes several key links: (1) seedsmen, (2) farmers, (3) consolidators to condition and handle, (4) a freight agent, and (5) an independent, third-party verification agent.

Soybeans Differ

The buyer's challenge is to find optimal soybeans for the particular application. About 20,000 soybean cultivars can be found in the U.S. soy germ plasm collection at the University of Illinois. The Chinese collection contains several thousand more. Figure 13.2 shows some of the variation in soybeans

FIGURE 13.2
Soybeans come in an array of sizes, shapes, colors, and qualities. Shown is a small sample of the varieties found in the soy germ plasm collection at the University of Illinois. (Photograph by Kelly Huff.)

held in the U.S. collection. These varieties differ significantly in size, shape, color, protein, oil, process characteristics, and flavor. Size, for instance, ranges from 600 soybeans per pound to over 8000 per pound; color varies from red to yellow to white to black; and protein content ranges from 34 to 58%. Commercial varieties in the United States come from fewer than 10% of these cultivars. Scientists continue to study this tremendous array of genetic material for potential benefits. Those wishing to conduct their own studies can secure small quantities of seed for multiplication.

Commodity Beans

Commodity soybeans sell by grade standards. Sellers win by delivering the lowest possible quality required within the buyers' grade specifications. Grade standards focus on a few physical features such as test weight, heat damage, total damage, splits, other colors, and the presence of stones. Such factors have little to do with the value of the soybean for a particular food application. They give no clue as to the biochemical characteristics that determine the success or failure of processing plans and foods. Figure 13.3 shows commodity soybeans being loaded onto barges at Asuncion, Paraguay.

FIGURE 13.3
Soybeans from Paraguay being loaded onto a barge at Asuncion. These are commodity soybeans sold to some grade standard. The barge will contain many varieties, all valued on crude physical properties and undifferentiated for food applications.

Variety Holds the Key

In selecting soybeans, choice of variety reigns. It is the single most important factor in determining suitability of supply. Variety probably accounts for 60 to 75% of the variation in value among soybeans. The place of production, weather, soil nutrition, and handling are other important variables. More than most, Japanese soyfood makers understand the importance of variety to the success of processing and producing good foods. They even appreciate that variety defines flavor. The best buyers can determine the suitability of a variety just by nibbling a representative soybean directly out of a production field.

Selection Guide

To make appropriate selections of soybeans for food applications, start with the market and work backward through processing to the supply and

FIGURE 13.4
A field of organic food soybeans growing on the Illinois prairie. A soymilk processor contracted for production of this preferred variety.

production of ingredients and raw materials and the development and supply of the seeds used to produce them. What is the product? Who is the target consumer? What characteristics does the consumer prefer? How can one distinguish this product from those of the competitors? Armed with such information, buyers can experiment to find optimal varieties. Knowledgeable suppliers can provide helpful guidance by suggesting a reasonable range of appropriate variety samples. For companies not having their own research and development laboratories with scaled processing, various universities offer confidential laboratory services. Some provide a full range of processing equipment, formulation services, and sophisticated consumer testing of products. With their diverse populations of foreign students, large U.S. universities can even provide quick cross-cultural testing. By using small batches, one can easily determine the range in process yield and end-product characteristics. These factors vary more with the variety available than with any other selection factor. Such experiments can define a particular variety or range of varieties that a particular processor might prefer — primarily for taste, process yield, and price. Figure 13.4 shows a field of soybeans being raised under contract for a maker of soymilk. That processor has identified this variety as providing the best flavor and process yield and has standardized the product based on this particular variety.

Consumer Values

Traditionally, consumers have made purchase decisions on taste, price, nutrition, color, and consistency. Recently, some consumers have begun to pay attention to more detailed social values, even investigating the technology used to develop the seed used to produce raw materials, the production culture used to grow the soybeans, and the materials and techniques used by the processor. Such "green values" support choices such as the following:

- *Conventional production* — More applicable to the second half of the 20th century, conventional agriculture tills or direct-seeds soybeans and controls weeds, insects, and molds with pesticides. Production decisions are disassociated from direct consumer choice. The big advantage of the conventional system lies in the production of huge volumes of commodity soybeans at very competitive prices.

- *Organic production* — In North America, Europe, and Japan, organic markets have been growing rapidly. In the United States, the rate of growth has exceeded 20% for the past 15 years. Organic consumers are willing to pay significant premiums for foods prepared in compliance with the rules of organic certification. The organic community opposes the use of synthetic pesticides, manufactured fertilizers, and genetically engineered seeds. It supports crop rotation, cover crops, and the use of appropriate cultivation. To market a premium product to this community, a processor needs to develop a secure supply chain that meets organic requirements. Because physical and chemical testing cannot verify compliance with cultural factors, suppliers have developed systems of process certification supplemented where appropriate by test verification and audits. Such market factors consistently encourage buyers to use identity-preserved supply systems that offer traceability back to the producer and greater food security. So, beginning with the consumer, the buyer of organic soybeans for food applications must consider factors throughout the supply chain from seed supplier to farm to process plant and the logistics that tie them together.

- *Pesticide-restricted production* — Some buyers offer a premium price to avoid the use of particular chemicals. Japan has for some time supported a premium market for pesticide-restricted soybeans.

- *Traceable supplies* — Consumers in Asia, North America, and Europe have recently become much more interested in knowing more about their food supply, including where and how the raw materials used to make their food were produced. To answer such questions, the processor must use a supply system that can trace production back to a particular group, farm, or field.

- *Selected genetically engineered features* — Some buyers may want a particular soybean trait available only in certain varieties. They may be looking for a particular functionality or health aspect in the oil or protein. While such traits might be measured and extracted from a commodity flow, the results correlate most directly to the selection of variety.

- *Avoidance of genetically engineered features* — Some buyers may wish to avoid as much as possible any contact with genetically engineered crops or a particular trait found in such crops. Complete avoidance has become realistically impossible. Buyers may establish tolerance standards and require testing for compliance. Without going into the details of regulation, the following reasonably describes current commercial limitations. The Europeans apply a formal tolerance standard of 0.9% with the requirement that soybeans or soybean products having more must be labeled as such. The United States has no official standard but many private standards, with those standards tending to match the European standards. The Japanese have an official standard of 5% but a widely accepted commercial standard of 1.0%. A survey of the world reveals a host of uncoordinated standards regarding the labeling and acceptance of genetically engineered materials. For a processor seeking optimal market flexibility, such standards become critical. While the press pays attention to official standards, food processors pay even more critical attention to informal standards. The tightest in the world of regular commerce at the moment are those imposed by the major grocery chains in the United Kingdom. They require labeling or rejection if materials carry more than 0.1% genetically engineered materials. For those willing to forgo marketing in the United Kingdom, the most common commercial target today is 0.9%.

Processor Values

When selecting optimal varieties, buyers need to address such process factors as the following:

- *Homogeneity* — Variations in supply quality trouble processors. Such variations require changes in recipe, temperature, dwell time, and formulation and make product standardization difficult. By using optimal varieties, especially ones grown in a particular area, buyers can optimize homogeneity and the collateral benefits of improved yield, fewer adjustments of recipes, and greater consistency in food products.

- *Market access* — Ideally, a processor would seek maximum market flexibility. That would suggest adopting the tightest known informal market standards for the presence of genetically engineered materials. It would also require selecting raw materials and ingredients required by any targeted market niche.

- *Process yield* — Process yield and efficiency vary greatly with variety selection. Testing will show which varieties and characteristics contribute to greater throughput. Selection of variety can remarkably reduce process time and increase the rate and quality of output. If water absorption is important in the process, a preferred soy characteristic might be quicker absorption. Other key selection criteria would involve sugar relationships, flavor, type of oil, and percentage protein. Variety choice is critical to each.

- *Supply security* — Having committed to a particular variety or quality of soybeans, buyers must arrange to receive what they need when they need it. Because the commodity system focuses overwhelmingly on yield per hectare, it does not offer selected characteristics nor does it offer the opportunity to buy on spot markets. To get farmers to produce what they want, buyers must contract with them early enough before planting so they can obtain the appropriate seed. Planting typically occurs in the northern hemisphere in May; in the southern hemisphere, in November. Buyers typically contract for a full year's supply to be delivered on scattered shipment over the time frame important to them. To ensure a sufficient supply in case of drought or other weather or disease problems, supply managers typically arrange for the farmers to produce 130% of what they need. Should that result in a surplus, it would be carried over as a rolling surplus into the next year or sold to accommodate another processor.

- *Price* — Price is always a consideration. The lowest priced bean is often more expensive in overall costs than a more expensive bean. The lowest priced bean may suffer in terms of process yield or water absorption and process efficiency. It might even preclude market acceptance. Nevertheless, buyers have a valid interest in obtaining a competitive price on their combined service/product supply arrangements. Clearly, price acceptance varies with the market. For example, the same Vinton 81 soybeans would vary in price depending on the receiving market. The following prices are representative of levels during early 2005:
  - U.S. #1 commodity market — $6.00/bu or $220.40/metric ton
  - Non-genetically modified U.S. #1 commodity market — $6.50/bu or $238.77/metric ton
  - Organically certified U.S. #1 for tofu — $18/bu or $661.20/metric ton

- Organically certified Vinton 81 soybean — $23/bu or $844.87/ metric ton

Price is directly linked to farm yield, ease of harvest, and production; all the normal rules of supply and demand; and the supply and service attributes required by the buyer.

- *Service* — The processing plant requires a supply system that will work with it to deliver consistent quality on demand with scattered shipment. Such quality requirements could be for cleaned, sized, and ready-to-cook product or merely bin run, depending on the needs and abilities of the buyer. Numerous companies in Asia, North America, and South America specialize in providing such services. For meeting specialty Japanese markets, the South American suppliers must contend with the disadvantage of always shipping product through an equatorial hot zone. The most particular Japanese clients avoid risking such high temperatures, which might contribute to oxidation and a consequent change in flavor. Such buyers take their entire supply for the year prior to the warmest months. What is not used immediately goes into chilled storage until needed.

Farmer Values

The farmer's basic interest lies in securing an appropriate combination of field yield and unit price to maximize his net return per acre. Yield per acre for varieties used for such purposes as soymilk, tofu, natto, or flour production might reasonably range in yield from 25 to 60 bushels per acre. To produce and identity preserve a high-yielding soybean might cost as little as $0.50/bu or $18.40/metric ton over the cost of a commodity soybean. The premium for a good conventionally produced tofu soybean with reasonable yield potential might run $2/bu or $73.47/metric ton. Production under organic certification would likely more than double the delivered price, although that ratio will likely fall over time. The farmer's secondary interest lies in prompt delivery. He would like to convert his product to cash as soon as possible. If the production contract requires the farmer to store the soybeans on the farm for a lengthy period, he will seek a storage premium to cover storage risks and the time value of money. In selecting farmers for specialty production, infrastructure is important. It is especially important that the farmer use a combine that treats the soybeans gently, typically a rotary combine. It is also important that the farmer provides storage in which the soybeans can be protected and gently handled. It would be best if the farmer harvested the soybeans at about 16% moisture and then dried them with ambient air to the contract specified moisture. But, it is even more important that the farmer display an attitude of service to the client, an understanding that he is producing a food product

FIGURE 13.5
Allen Williams, an organic farmer, inspects the plant health of organic food soybeans.

for which quality is important to the processor and ultimately the consumer. That requires a farmer selection process as well as a variety selection process. Figure 13.5 shows an established organic farmer checking soybean plant quality. The buyer prefers this particular variety for flavor and contracts with this farmer each year to produce them.

Seed Suppliers

Seed suppliers, especially smaller regional suppliers, show a keen interest in developing soybeans for specific food applications. Even in countries such as the United States that have shifted overwhelmingly to genetically engineered soybeans, seedsmen continue to supply traditionally bred soybeans for most popular food applications. Several companies now specialize in screening all the commercial germ plasm collections to find and offer those of greatest value for food applications. Processors who would like to use a new variety or develop their own variety should understand that 7 years of development are required before it is known if the new variety really performs to the stated requirements. The quickest path to choice is to investigate new and existing commercial varieties developed by breeders hoping to capture the attention of food processors.

Supply Systems Compared

The commodity system treats all soybeans as fungible. It sorts and delivers soybeans by grade standard, not by any process criteria. The main benefit of the commodity system is that it delivers huge volumes of raw material very efficiently. It carries a nearly constant surplus and permits buyers to make spot purchases at almost any time, but it does not segregate soybeans by cultural practices or variety. For those, buyers need to work with the segregated systems mentioned above that preserve the identity of the chosen soybeans from farm to processor. Identity-preserved systems generally depend on contract production and segregated storage, conditioning, and logistics support. Farmers typically store the contracted crop in farm storage, releasing it on the call of the associated merchandiser from the time of harvest to about 2 months before the subsequent harvest. At that time, the merchandiser typically takes enough of the segregated crop into commercial storage to clear the farm infrastructure to receive the incoming harvest. For the buyer, the disadvantages of this system are that needs must be anticipated perhaps a year before harvest, there is no spot market, and it is not possible to obtain the particular variety should a crop shortfall occur. The advantage is that buyers get what they want when they want it.

Logistics Comparison

Soybeans shipped through the commodity system normally travel in bulk carriers, with a single barge carrying 1350 metric tons, a grain train 8000 metric tons, and a Panamax vessel 50,000 metric tons. Identity-preserved crops often travel in bulk or bag by containers carrying no more than 20 to 24 metric tons. Containers typically travel door-to-door, move individually or in groups of five, and require a minimal cash outlay by buyers. Unlike bulk commodity carriers, containers facilitate traceability and require minimal handling of products. In some situations, the 20-ton ocean freight containers offer better freight rates than do bulk commodity carriers. Some of the most favorable rates are found where trade imbalances exist. For example, containers arrive in the United States from Asia carrying a vast array of consumer products. Because of the lack of corresponding trade to carry high-value materials from the United States back to Asia, container companies offer very attractive rates for the return trip, using the western trip to subsidize the return freight. Soybeans can be shipped in containers in paper bags typically weighing 30 kg, in 1-ton totes, or in bulk behind wood or cardboard barriers.

FIGURE 13.6
Typical farm storage units in the Midwest. Each bin holds about 435 metric tons of soybeans and has a perforated aeration floor with fans hooked to a computerized controller that samples weather conditions and holds soybeans at desired moistures and temperatures.

Contract Manager

Buyers sometimes arrange all the aspects discussed above by themselves. More typically, buyers contract with a single company located in the primary production zone to assist in selecting the appropriate variety; to arrange seed supplies and production contracts; to supervise production and keep the buyer informed of crop conditions; to receive, prepare, package, and ship the soybeans after harvest on the buyer's schedule; to contract freight; and to manage the entire supply chain. Such relations range from short to very long term and are governed by the normal rules of competition and client satisfaction. Such supply chain managers often balance competing values. For example, for maximum homogeneity, it would be preferable to buy the selected varieties grown in close proximity; however, for production security, it would be desirable to produce them over widely scattered terrain to minimize the risk of weather, local disease, or insect problems. Farm storage is the ideal system for scattering risk and gently handling products, but concentrated commercial storage might give better supply control. Fortunately, soybeans present few serious storage problems because of the diversified farm storage units across the U.S. Midwest and the common storage bag systems found in South America. (See Figure 13.6 and Figure 13.7.)

Third-Party Verification

If the buyer focuses strictly on characteristics that can be easily measured and tested, supply-chain verification remains simple and straightforward. But, if the buyer is interested in markets that are sensitive to the presence of transgenic materials or want their foods to come from production that

FIGURE 13.7
Ag bag on the Argentine pampas. The use of heavy, well-made plastic bags to store soybeans has soared in Argentina. Standard bags hold about 200 metric tons and allow farmers to store soybeans for months to avoid moving them to market at harvest. The system works well for segregating varieties.

meets organic or other cultural standards, verification becomes more critical and much more difficult.

Transgenic Sensitivity

With regard to receiving or avoiding a particular transgenic trait, seed supply is critical to meeting accepted tolerances. If the client's tolerance level is 0.9%, the seed must fall clearly within that tolerance. The presence of transgenic materials becomes a more critical issue along the path from seed to consumer. The supply chain is fraught with multiple opportunities for contamination. The seed itself might be contaminated. Compliance cannot be verified by visual inspection. Assuming the seed is within tolerance when delivered, the farmer might compromise the quality by improperly mixing the seed or planting it in the wrong field. After it has been planted, the field might be contaminated by cross-pollination with soybeans that do not meet the standards. Although soybeans self-pollinate, insects often insinuate themselves in the process. Few insects respect field borders. At harvest, seeds from other noncompliant fields might be mixed during the harvest process or by the presence of other soybeans in trucks moving the harvest into storage. The storage could be contaminated by error if the farmer is not paying seriously detailed attention. Post-harvest contamination can occur when the soybeans are moved from storage to the delivery vehicle, when other beans are in the vehicle, when mistakes are made at the receiving facility, or during cleaning, conditioning, packaging, or shipping. Contamination can occur from dust from poorly cleaned truck tarps or nearby activity involving other soybeans. Such contamination would not be obvious on visual inspection. It would require serious efforts at taking representative samples and then thoroughly testing those representative samples by biochemical assay, DNA study, or protein testing.

While supply-chain managers can perform appropriate sampling, inspection, and testing, they become commercially interested parties. For that reason, most good suppliers recommend that "process" and "product" be verified by independent, third-party inspectors whose business relies completely on their reputation for integrity and accuracy. Several reputable companies offer such services in North America, South America, and Asia. Sometimes they combine in one entity the sampling and testing functions, and sometimes such companies contract one side of the service to another party. The U.S. Department of Agriculture's Grain Inspection, Packers, and Stockyards Administration (GIPSA) conducts well-defined ring tests with standardized transgenic samples to check the accuracies of testing laboratories serving such organizations. It publishes the results of those tests or at least the favorable results of those tests. So, readers looking for accuracy in genetically modified organism (GMO) testing might want to check the GIPSA website.

Even with the best testing services, consistency remains a difficult challenge. There is no guarantee that two tests will produce exactly the same results. If they differ, which test applies? Buyer and seller need to stipulate the definitive test in their contract. Costs to the system would be significantly lowered if point and time of shipment testing ruled, but, until governments agree to reciprocity or to standardized testing accepted by all, that is often impossible. Moreover, sampling itself still remains a major challenge even after all parties have agreed to testing protocols. Whose sample is representative? What is the value if it is not representative? What size sample is needed for appropriate statistical accuracy? How does one take a "representative" sample?

Process Verification

The problem grows as buyers shift their focus from factors that can be tested (transgenic materials) to factors that cannot be tested, such as the organic nature of raw materials, ingredients, and foods. In the United States, *organic* is defined as a process, not a condition. Soybeans and the consequent soy foods can be certified organic if they are produced in compliance with national organic standards. With slight exception, those standards do not physically define organic. So, for example, the certification of soymilk as organic depends on verification of the production process going back to the seed supply and working through production, supply, cleaning, processing, packaging, and delivery. Testing is neither determinative nor stipulated. Instead, the United States has created a system requiring third-party "organic certification" by inspectors approved by the USDA's National Organic Program (NOP). Each link in the chain must be visited, process audited, and inspected annually to ensure compliance with the organic rules. By early 2005, roughly 100 companies and individuals scattered over many countries had been approved to certify products to the USDA organic standards. USDA inspectors are active under this program, investigating approved certifiers in China as well as North and South America and Europe. Maintaining consistency remains a management challenge.

Layered Sampling and Testing

Some markets are extremely sensitive to the presence of transgenic materials. Companies most concerned about perfection in such supply situations have developed verification systems that rely on multiple, layered sampling and testing. Fresh Pure Green™, a Japanese program, stipulates the following sampling protocols to meet acceptable transgenic tolerances:

- Seed sampling
- Visual inspection of the production field when leaves are dropping
- Sample from combine at harvest
- Sample from storage bin shortly after harvest
- Sampling of all shipments before unloading at the consolidators where the soybeans are to be cleaned, sized, and packaged
- Random sampling of the inventory and packaged product at the consolidator's plant

With the exception of the sample delivered to the consolidator's plant, all samples are subjected to chemically indexed germination tests, an enzyme-linked immunosorbent assay (ELISA) strip test, an ELISA bench test, and a DNA test. The sample being delivered to the consolidator comes from an approved bin and, under time pressure, is tested with an ELISA strip test. When production plots have been well designed and the identity-preserved program is well disciplined, such verification protocols generally reject less than 3% of the production as failing to meet tolerances — even in the United States, where transgenic materials dominate.

Summary

Done well, the full identity-preserved system achieves the worthy goal of making a client happy. Selection of the soybean itself starts with the consumer and ties most directly to the choice of varieties. Most companies experiment with 5 to 20 varieties to select their preferred soybeans. This selection is complemented by the choice of an appropriate production zone, but the result again ties to the market and goes well beyond a narrow focus on the soybean itself. The best buyers contract with a supply-chain manager and create programs that select producing farmers by location, infrastructure, and attitude. To avoid error and commercial embarrassment, the contract manager employs a variety of sampling and testing protocols to accompany all the standard rules of quality control. The result is happy clients who get what they want when they want it and products that are traceable back to the farm and farmer. The following factors will continue to encourage processors and retailers to move toward identity-preserved supply systems:

- *Genetic engineering* — Genetic engineering will become an even more important player as dozens of new genetic traits make verification exponentially more difficult.
- *Affluent markets* — Increasingly, affluent consumers want food raised according to their social values. While organic production may never dominate, the organic market will be a very important niche in Europe and North America and probably in Asia.
- *Ability to measure* — Distinctions from both social and biological perspectives increase with improved ability to measure. New techniques and more sophisticated versions of existing techniques and equipment are becoming available.
- *Storage flexibility and isolation techniques* — U.S. and South American experiences over the past decade demonstrate the excellent ability to segregate and successfully deliver non-GMO or particular GMO materials within very tight tolerances.
- *Process yield experience* — Processors are learning that the hidden cost of the commodity system lies in reduced process yield as well as product variation. Identity-preserved systems can increase process yield well above 10% and sometimes above 30%.
- *Traceability* — As food scares fan public concern in Asia, Europe, and North America, more consumers want to know the source and path of their food. The identity-preserved systems that provide that security lend themselves to the sourcing of excellent soybeans.

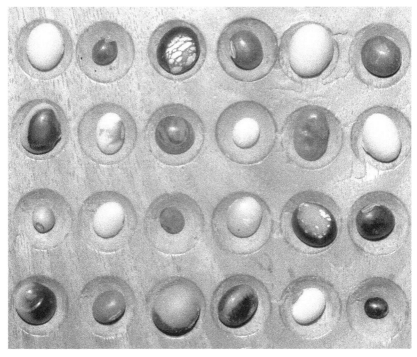

FIGURE 13.8
Varieties differ. This inspection tray displays some of the physical features of varieties held in the U.S. soy germ plasm collection. (Photograph by Kelly Huff.)

Soybeans are not created equal in terms of the characteristics that matter to making foods. They differ significantly in terms of physical and biochemical properties. As Figure 13.8 shows, many choices are available. Diligence in choosing provides more than its own reward.

14

World Initiative for Soy in Human Health

Jim Hershey

CONTENTS

Introduction

The World Initiative for Soy in Human Health (WISHH) program began as a simple concept articulated by the Illinois Soybean Producers' Operating Board (ISPOB) which was to *increase the amount of soy protein consumed by people in developing countries*. Economic modeling done by the University of Illinois confirmed what the soybean farmers intuited: With even moderate economic and population growth, many parts of the world, and many people in even wider areas, will suffer from shortages in protein for decades to come. As concerned world citizens and savvy producers of the product needed to satisfy that demand, U.S. soybean farmers joined the Illinois leaders and formed the International Soy Protein Program (ISPP).

A New Program Is Born

Soybean farmers in the United States knew from decades of experience that investments in long-term, demand-building, generic promotions can create and expand markets. Farmers in the United States have a tradition of generating funds for promotion and research through a system called *checkoffs*, which require farmers of a given commodity to pay a percentage (or sometimes a fixed amount) of their commodity sales into a fund managed by a board of farmers. In the case of the national soybean checkoff, half of the funds generated stays in the states of collection, and the other half goes to the national entity called the United Soybean Board.

While most checkoff-funded activities are approved on an annual basis, the ISPOB guaranteed a 3-year startup period and budget of $1 million per year. A number of other state checkoff boards recognized the power and potential of this idea and chose to share the expense of the program with the Illinois board. The following states joined Illinois as first-year contributors to the WISHH program: Indiana, Iowa, Michigan, Minnesota, Missouri, Nebraska, Ohio, South Dakota, and Wisconsin. By the end of the third year of the WISHH program, 15 state boards and the national body, the United Soybean Board, were contributing regularly to WISHH.

The ISPOB and Executive Director Lyle Roberts helped conceive of the program and raise the funds, but they also had to consider how the program would be organized and managed. Because the funds were new and the concept of a program devoted only to soy protein in food was an innovation, the farmers and staff felt that synergies and economies of scale could be achieved by having the program managed by the American Soybean Association (ASA). At the July 2000 meetings of the American Soybean Association and United Soybean Board, the ISPOB and the ASA formalized their commitment and set this objective:

> Increase the international consumption of soy protein by humans in new markets — developing countries — and thereby create new opportunities for the consumption of soybeans and provide greater economic returns to U.S. soybean producers.

The International Soy Protein Program (ISPP) was underway.

The First Year

The ASA's International Marketing Program had been handed an opportunity, but it was a responsibility, as well, so Jim Guinn, ASA International Marketing executive director, set about finding a director to head the ISPP.

An 8-year ASA International Marketing veteran, with an accumulated 15 years of U.S. agricultural commodity market development experience, Jim Hershey understood the challenges and opportunities the new program presented and was appointed director of the newly funded soy protein for human nutrition program on November 1, 2000.

The ISPOB and Jim Guinn, with help from the University of Illinois' National Soybean Research Laboratory, outlined some areas of activity and project development and targeted organizations distributing food aid and working on devastating diseases. Hershey built on those concepts and pulled together the working group for a strategic planning meeting in December 2000. One item on the agenda was to create a name for the program and, in less than an hour, the name *World Initiative for Soy in Human Health* (WISHH) was chosen. The group took longer to identify key target areas and strategies, but by the end of the day the direction was set.

The WISHH team, made up of WISHH, ASA, and ISPOB staff, as well as The National Soybean Research Laboratory and consultants, picked the following strategic areas on which to focus resources to increase the varieties and volumes of soy proteins moving into international channels:

- Soy food aid/humanitarian efforts
- Private voluntary organization (PVO) planning
- World Food Programme (WFP) assistance
- Global Food for Education Initiative
- HIV/AIDS initiative

A heavy emphasis was placed on opportunities in food aid, and one of the early endeavors was to get value-added soy proteins registered for procurement by the U.S. government for international programs.

In January 2001, the USDA's Farm Services Agency issued its official document, *Value-Added Soy Proteins 1* (VASP1), complete with specifications and other details to guide sellers on selling soy proteins to food aid programs. The VASP1 included the following products:

- Defatted soy flour (DSF)
- Textured soy protein (TSP)
- Soy protein concentrate (SPC)
- Isolated soy protein (IPC)
- Dried soymilk powder (DSP)

By the end of the first year, the USDA had tendered for 2500 metric tons of DSF bound for Indonesia for a noodle fortification project, inspired by work done originally by the ASA Singapore office. This initial sale demonstrated that food aid could move important volumes of new (to food aid) value-added soy products for human nutrition.

The WISHH program also spent much of the first year gearing up its technical assistance programs to demonstrate how VASPs could function and be accepted in feeding programs around the world. Generally, WISHH reached out to PVOs and the World Food Programme, seeking partners for pilot sample demonstrations. One early and eager partner was Africare, who was interested not only in VASPs for nutritional interventions but also in soybean processing for small enterprise support. With this motivation, Africare committed to purchase 14 "soy cows," boiler and grinding machines of Canadian design. With WISHH's guarantee of technical support and startup training, Africare placed the machines in 7 countries and reports of successful and profitable operations continue to this day.

Another activity WISHH undertook with Africare was the shipment of an entire container of soy protein to their feeding programs in Angola. This container and several others destined for other countries and other PVOs in Africa were graciously donated by Archer Daniels Midland (ADM). WISHH consultants could not travel to Angola because of the civil war that was winding down at the time, so the Africare project leader was trained at another training program that WISHH was conducting in Zimbabwe. These programs, the small- and large-scale samples, and the technical training typify the WISHH strategic approach. WISHH works with partners to demonstrate: (1) the applicability of soy to their programs and needs, and (2) acceptability by the recipients who become the consumers of the product.

Africare personnel also played *de facto* matchmakers in 2001 when they introduced WISHH to an organization called Humana People to People, an international nongovernmental organization (NGO) with offices and partners in many countries. Humana had a program in Zimbabwe where they were using soy protein to bolster the health and immune systems of citizens, HIV positive or not. Humana approached the WISHH program seeking donations of soy protein for their Total Control of the Epidemic (TCE) program. WISHH, with another gracious donation of TVP® from ADM, shipped approximately 10 tons to Humana in Zimbabwe for distribution through TCE's innovative soy restaurants, which are small enterprises serving as educational centers to teach the population about HIV and as feeding sites for vulnerable populations such as orphans and families affected by the deadly virus. Collaborations with PVOs and activities addressing the nutritional challenges presented by the HIV/AIDS pandemic were poised to increase the usage of soy protein in people's diets.

Years Two and Three

The years 2002 and 2003 witnessed the continued ramp up of WISHH activities in developing countries, including memoranda of understanding with PVOs and corresponding sampling activities in Honduras, Kenya, Senegal,

Tajikistan, and Uganda. Another feature that emerged was the U.S. soy processing industry's willingness to donate not only samples but also cash resources to conduct training and outreach programs. ADM and Cargill agreed to fund technical activities conducted by WISHH in target countries; a project in Honduras was supported by Cargill, and a pilot project in Cote d'Ivoire was supported by ADM. While the programs differed somewhat, one factor the two projects had in common was to introduce textured soy protein (TSP) to local school lunch programs. In fact, this has been one of WISHH's key strategic focus areas. Just as the U.S. school lunch program and American school cooks utilize TSP to their advantage, feeding program managers also appreciate the virtues of TSP. These attributes include:

- Comprised of 50 to 55% protein, TSP is one of the most protein-dense products available.
- TSP has a long shelf-life, a minimum of 1 year, with anecdotal field experience suggesting an even longer shelf life.
- TSP is quick cooking; because the product is precooked, it can be added at any stage of food preparation.
- Versatile and acceptable, the product soaks up the flavor of the sauce or soup to which it is added, making it more acceptable to local consumers, particularly children. The fact that the TSP approximates meat in its visual and tactile appearance increases its acceptability.

In its pilot sampling programs, WISHH measured product acceptability with both cooks and consumers and found that product acceptability uniformly scored greater than 85%. TSP is a popular choice when PVOs seek to program protein products for their food aid requests. By 2005, TSP had been included in six food aid programs authorized by the USDA, for a total of 3240 metric tons. WISHH is working to increase those figures and to see TSP moved through the largest U.S. food aid program (by dollar value and volume), the U.S. Agency for International Development (USAID) Food for Peace Title II program. In fact, in 2004 USAID approved two Food for Peace programs for VASPs.

2004 and the WISHH Strategic Plan

At the end of fiscal year 2003, it was time for WISHH to take stock. Its 3-year startup period was over. The WISHH Advisory Panel, a group of farmers and staff from state checkoff boards and ASA and United Soybean Board (USB) farmer board members, directed WISHH to undertake an evaluation and strategic planning process. The WISHH team knew that in order to maintain the project's implementation and expansion, it would be necessary to find an outside facilitator. At the same time, the value of the eventual plan would be directly proportional to the effort and engagement the team brought to the strategic planning process.

The WISHH team (staff, consultants, and volunteer panel members) delivered in-depth information and thoughtful analyses. The contracted facilitator organized the surveys and focus groups, collected and correlated the data, and wrote a series of documents and drafts that captured where WISHH had come from, what it had learned, and, upon digesting this information from the environmental review, where WISHH should go. Key to the plan were the strategic areas and objectives, both short- and long-term, that will move WISHH toward fulfilling its mission. In fact, the mission statement itself was a newly crafted declaration of a familiar theme:

> To create sustainable solutions for the protein demands of people in developing countries through the introduction and use of U.S. soy products.

The key phrase in this statement is "sustainable solutions," which, according to the WISHH team, means commercial activities involving soy protein. Everyone, from recipients to providers, recognizes and believes that food aid is only a short-term solution. The way to address the long-term nutritional and protein needs of people is to ensure that such products are available on the market in forms that people will want to buy and at prices they can afford. So WISHH activities are geared increasingly toward commercial market development.

Other organizations also share WISHH's view of sustainable solutions, and WISHH is teaming up with them to increase the utilization of soy proteins, especially in commercial applications. Private companies, USAID, and the USDA have all agreed to fund various projects, providing WISHH with powerful leverage for its soybean checkoff resources. This situation combines two of the key strategic areas in the new WISHH plan: commodity utilization and leveraged funding.

Into the Future

Market development, through demand-building activities and technical transfer, is a long-term process. U.S. soybean farmers have a long tradition of taking the long view and assuming considerable risk to create interest for U.S. soy. Many of their bets have paid off over the years; the most dramatic, though not the only, example perhaps is China, which has moved from a net exporter when ASA opened their China office in the 1970s to the single largest importer of U.S. soybeans — over 8 million metric tons in 2004. While human consumption of soy protein might not generate figures like that, U.S.

farmers know what they need to do to accelerate the adoption of soy consumption by consumers and to position the United States as the premier supplier of choice within a generation.

Index

A

acesulfame-K, 190
acetyl glycosides, 26
acid, 230; *see also* ready-to-drink acid
acid leaching, 52, 208
acid treatment, 239, 240–241
Acidophilus, 15
additives, 145
Africa, use of soybeans in, 6–7
Africare, 270
aglucones, 118
air, trapped, 95
albumins, 206, 236
alcohol removal, 242
aldehydes, 231, 233, 237, 242, 243
alfalfa leaf protein concentrate, 236
alitame, 190
alkanals/alkenels, 233
allergies to soyfoods, 70
alpha-linolenic acid, 10
American Association of Cereal
 Chemists (AACC), 65
American Soybean Association
 (ASA), 268, 269, 271
amino acids, 40, 73, 78, 94, 202, 211,
 229, 235, 243
 in soybeans, 10
amylose, 104
animal feed, 4
anticarcinogens, in soy, 28; *see also*
 cancer
antioxidants, 31, 233
 in soybeans, 11
antivitamins, 230
aqueous alcohol extraction, 208
Argentina, 3

arginine, 94
ascorbic acid, 75
Asia, use of soybeans in, 5–6, 27
aspartame, 190
aspergillopeptidase, 243
Aspergillus oryzae, 16
autoxidation, 230–231, 239

B

Bacillus natto, 16
baked bars, 188
baking, soy ingredients and, 63–79,
 230
 applications of, 71–79
 functions of, 65–70
 health benefits of, 70
 types of, 64–65
bar hardening, 197
barbeque sauce, labeling of, 167–168
beanyness, 115, 116, 244
beef patties, 164, 166, 172–175
Belgium, 6
beta-carotene, 11
beta-conglycinin, 41
Bifida, 11, 15
binders, 139, 169, 181
biotin, 11
biscuits, 77–78
bitter peptides, 234–235
bitterness, 115, 116, 117, 118, 120, 121,
 243, 244
 masking, 245
blanching, soybean, 120–121, 122,
 125
bleaching, 70
bone mineral density, 30

shaping, added protein and, 87
shear, 210, 211, 222
shelf life
    of bakery products, 70, 77, 78
    of food bar, 196–197
    of nutritional beverages, 201,
        202, 216, 217, 223
shoyu, 16
Silk® brand soymilk, 8
smoothies, 205, 221, 228
snacks, protein-enhanced, 83–92
soaking, 120–121
sodium, 11, 241
sodium alginates, 213
sodium bicarbonate, 121, 194
sodium carboxymethylcellulose
    (CMC), 222
sodium caseinate, 195
sodium stearoyl-2-lactylate (SSL),
    74, 78, 220
sodium stearyl fumarete (SSF), 78
solubility, 66–68, 240
soluble soy protein stabilizer (SSPS),
    125
sorbitol, 190, 193
soy allergies, 70
soy base
    extract, 111–132
    extraction equipment suppliers,
        126–128
    extraction process, 114–129
    separation, 123
soy-based infant formula (SBIF), see
    infant formula
soy, blending with semolina, 99–100
soy bran, 58, 64, 65, 78
soy burgers, 229
soy cell wall, 59
soy concentrates, see soy protein
    concentrate
soy crisps, 85
soy crumbles, 157
soy fiber, 42, 58–61, 94, 125; see also
    fiber: in soybeans
    cotyledon, 59–61
    hull, 58–59

soy flakes, 60, 157, 208
    defatted, 143, 144, 206
    enzyme-active, defatted, 41,
        46–48
    enzyme-inactive, defatted, 41,
        48–49
    full-fat, 142
soy flour, 3, 7, 12, 20, 40, 41, 64, 156,
        157, 191, 206, 208, 229, 230,
        243; see also baking: soy
        ingredients, and
    breadmaking, and, 74
    color control, and, 70
    defatted, 19, 64, 73, 74, 76, 77, 78,
        130, 138, 142, 159, 206,
        233, 269
        as extender in breads, 75
    enzyme-active, 74, 76
        color control, and, 70
        defatted, 41, 46–48, 64
        full-fat, 41, 42–44
    enzyme-inactive
        defatted, 48–49, 76
        full-fat, 41, 44–45
        low-fat, 41, 45–46
    flavor, 66
    full-fat, 10, 12, 18–19, 64, 75, 76,
        78, 112, 113, 130, 142,
        208, 240
        water-binding capacity,
            and, 69
    gels, 69
    high-fat, 64, 75
    lecithinated, 41, 50–51, 64, 76, 77
    low-fat, 64, 78
    mechanically expressed, 138,
        142, 144
    organic, 42, 62, 102
    powdered, 168
    refatted, 41, 51
    textured, 19, 41, 50, 159–160,
        164, 168
    water-holding capacity, and, 69
    whole grain, 65
soy germ, 25, 42, 55, 58, 64, 65, 118,
        120

soybean oil, *see* soy oil
soybeans
    aqueous extract of, 206
    cleaning, 42–43
    commodity, 251
    conditioning, 46–47
    cracking and dehulling, 43–44
    cultivars, 250
    dehulling, *see* dehulling
    drying, 43
    flake extraction, 47–48
    hemicelluloses, 222
    identity-preserved (IP), 13, 118,
        119
    milling, 44
    nutritional components, 9–12,
        94, 142, 206, 229–230
    per capita consumption, 5
    preparation, 119
    processing, 39–62
    production, 228
        worldwide, 4, 142
    selection, 118–119, 249–265
    selection guide, 252–253
    shipment, 259
    soaking/blanching, 120–121, 125
    storage, 257, 259, 260, 264
    utilization, 4–9
    varieties, 252, 265
        differences in, 250–251
    wet grinding, 122
    whole dry, 12–13
    yield calculation, 126, 127
soyfoods, 12–20
    classes of, 71
    health effects of, 23–34, 40
    history of, 2–3
    safety concerns for, 33
soymilk, 2, 4, 6, 8, 10, 12, 13, 14, 20,
        40, 60, 111–132, 200, 204, 205,
        219, 228, 229, 240, 253, 269
    emulsification, 120
    flavor, 112
    off-flavors, 239
    powder, spray-dried, 130
    powders, 114

production, 112
reconstituted, 130
sales revenues in U.S., 200
Silk® brand, 8
whole-bean extract, and, 206
yogurt, 15
soynuts, 17
specific mechanical energy, 102
sponge-and-dough method, 74
spun protein isolates, 139
stabilization, 128–129, 221–222
stabilizers, 213
stachyose, 11, 230, 244
starch, 139, 140, 156, 162, 170, 177,
    181
    gelatinized, 99, 102, 103–106
    pregelatinized, 104–105
statin, soy protein and, 31
stearic acid, 230
storage
    of grains, 91
    of soybeans, 257, 259, 260, 264
structure builders, 99
structured meat analogs (SMAs), 138
sucralose, 190
sucrose, 11, 217, 236
sugar alcohols, 189, 190, 192–193
sugar, effect on browning, 212–213
sugars, in raw materials, 142
sulfur, 11, 146
supercritical carbon dioxide
        extraction, 241–242
suppliers, seed, 258
supply chain, 250, 260, 261, 262
supply-chain manager, 264
supply, security of, 256
supply systems, 259
Suzuyutaka, 232
sweating, 98
sweetness, 115

T
tamari, 16
tempeh, 6, 8, 10, 15, 40
temperature control, to prevent off-
        flavors, 239–240